JAN 25 2017

W9-BKV-084

DISCARDED

Praise for
THE VITAL QUESTION

"Magnificent."   —Oliver Morton, *Intelligent Life (Economist)*

"If I were a rich man, I would buy up the print run of this book and give a copy to every science undergraduate ahead of his or her first course in cell biology."   —Franklin M. Harold, *Microbe Magazine*

"Why is all complex life a result of ... one event? ... The answer that Lane gives, if it's correct, has profound implications."
                    —Tom Chivers, *BuzzFeed*

"What Lane is proposing, if correct, will be as important as the Copernican revolution."   —Peter Forbes, *Guardian* (UK)

"An essential book for anyone interested in the origin of life."
                    —Ravinder Kanda, *Times Higher Education* (UK)

"This terribly important book . . . is a manifesto for the future of basic biology."   —Adam Rutherford, *Guardian* (UK)

"Comes triumphantly close to cracking the secret of why life is the way it is, to a depth that would boggle any ancient philosopher's mind."
                    —Matt Ridley, *Times* (UK)

"This is a book of vast scope and ambition, brimming with bold and important ideas. . . . [A]n incredible, epic story."
                    —Michael Le Page, *New Scientist* (UK)

"Succeeds brilliantly. . . . I cannot recommend *The Vital Question* too highly. Lane's vivid descriptions and powerful reasoning will amaze and grip the reader."   —Caspar Henderson, *Sunday Telegraph* (UK)

MAXWELL MEMORIAL LIBRARY
14 GENESEE STREET
CAMILLUS, NEW YORK 13031

JAN 25 2017

DISCARD

"An exciting tale, thrillingly told. . . . [T]his is a potent book, one that not only brings you up to date with biology but also stuns you with the wonder of it all."   —Bryan Appleyard, *Sunday Times* (UK)

"The best summary of how actually abiogenesis would have occurred."
                              —PZ Myers, *Pharyngula (Science Blogs)*

MAXWELL MEMORIAL LIBRARY
14 GENESEE STREET
CAMILLUS, NEW YORK 13031

NICK LANE

THE VITAL QUESTION

ENERGY, EVOLUTION, AND THE
ORIGINS OF COMPLEX LIFE

CAMILLUS

W. W. NORTON & COMPANY

Independent Publishers Since 1923

New York • London

Copyright © 2015 by Nick Lane
First American Edition 2015

First published in Great Britain under the title: *The Vital Question:
Why Is Life the Way It Is?*

All rights reserved
Printed in the United States of America
First published as a Norton paperback 2016

For information about permission to reproduce selections from this book, write to
Permissions, W. W. Norton & Company, Inc.,
500 Fifth Avenue, New York, NY 10110

For information about special discounts for bulk purchases, please contact
W. W. Norton Special Sales at specialsales@wwnorton.com or 800-233-4830

Manufacturing by Quad Graphics, Fairfield, PA
Production manager: Anna Oler

Library of Congress Cataloging-in-Publication Data

Lane, Nick.
The vital question : energy, evolution, and the origins of complex life /
Nick Lane. — First American edition.
pages cm
Includes bibliographical references and index.
ISBN 978-0-393-08881-6 (hardcover)
1. Life—Origin. 2. Cells—Evolution. 3. Energy metabolism. I. Title.
QH325.L36 2015
576.8'3—dc23
2015004610

ISBN 978-0-393-35297-9 pbk.

W. W. Norton & Company, Inc.
500 Fifth Avenue, New York, N.Y. 10110

W. W. Norton & Company Ltd.
Castle House, 75/76 Wells Street, London W1T 3QT

2 3 4 5 6 7 8 9 0

For Ana
My inspiration and companion
On this magical journey

CONTENTS

INTRODUCTION:
WHY IS LIFE THE WAY IT IS?

There is a black hole at the heart of biology. Bluntly put, we do not know why life is the way it is. All complex life on earth shares a common ancestor, a cell that arose from simple bacterial progenitors on just one occasion in 4 billion years. Was this a freak accident, or did other 'experiments' in the evolution of complexity fail? We don't know. We do know that this common ancestor was already a very complex cell. It had more or less the same sophistication as one of your cells, and it passed this great complexity on not just to you and me but to all its descendants, from trees to bees. I challenge you to look at one of your own cells down a microscope and distinguish it from the cells of a mushroom. They are practically identical. I don't live much like a mushroom, so why are my cells so similar? It's not just that they look alike. All complex life shares an astonishing catalogue of elaborate traits, from sex to cell suicide to senescence, none of which is seen in a comparable form in bacteria. There is no agreement about why so many unique traits accumulated in that single ancestor, or why none of them shows any sign of evolving independently in bacteria. Why, if all of these traits arose by natural selection, in which each step offers some small advantage, did equivalent traits not arise on other occasions in various bacterial groups?

These questions highlight the peculiar evolutionary trajectory of life on earth. Life arose around half a billion years after the earth's formation, perhaps 4 billion years ago, but then got stuck at the bacterial level of complexity for more than 2 billion years, half the age of our planet. Indeed, bacteria have remained simple in their morphology (but not their biochemistry) throughout 4 billion years. In stark contrast, all morphologically complex organisms – all plants, animals, fungi, seaweeds and single-celled 'protists' such as amoeba – descend from that singular ancestor about 1.5–2 billion years ago. This ancestor was recognisably a 'modern' cell, with an exquisite internal structure and unprecedented molecular dynamism, all driven by sophisticated nanomachines encoded by thousands of new genes that are largely unknown in bacteria. There are no surviving evolutionary intermediates, no 'missing links' to give any indication of how or why these complex traits arose, just an unexplained void between the morphological simplicity of bacteria and the awesome complexity of everything else. An evolutionary black hole.

We spend billions of dollars a year on biomedical research, teasing out the answers to unimaginably complex questions about why we get ill. We know in enormous detail how genes and proteins relate to each other, how regulatory networks feed back into each other. We build elaborate mathematical models and design computer simulations to play out our projections. Yet we don't know how the parts evolved! How can we hope to understand disease if we have no idea *why* cells work the way they do? We can't understand society if we know nothing of its history; nor can we understand the workings of the cell if we don't know how it evolved. This isn't just a matter of practical importance. These are human questions about why we are here. What laws gave rise to the universe, the stars, the sun, the earth, and life itself? Will the same laws beget life elsewhere in the universe? Would alien life be anything like us? Such metaphysical questions lie at the heart of what makes us human. Some 350 years after the discovery of cells, we still don't know why life on earth is the way it is.

You might not have noticed that we don't know. It's not your fault. Text books and journals are full of information, but often fail to address these 'childlike' questions. The internet swamps us with all manner of indiscriminate facts, mixed with varying proportions of nonsense. But it's not merely a case of information overload. Few biologists are more than dimly aware of the black hole at the heart of their subject. Most work on other questions. The great majority study large organisms, particular groups of plants or animals. Relatively few work on microbes, and even fewer on the early evolution of cells. There's also a concern about creationists and intelligent design – to admit we don't know all the answers risks opening the door to naysayers, who deny that we have any meaningful knowledge of evolution. Of course we do. We know an awful lot. Hypotheses on the origins of life and the early evolution of cells must explain an encyclopedia of facts, conform to a straitjacket of knowledge, as well as predict unexpected relationships that can be tested empirically. We understand a great deal about natural selection and some of the more random processes that sculpt genomes. All these facts are consistent with the evolution of cells. But this same straitjacket of facts is precisely what raises the problem. We don't know why life took the peculiar course that it did.

Scientists are curious people, and if this problem were as stark as I'm suggesting, it would be well known. The fact is – it's far from obvious. The various competing answers are esoteric, and all but obscure the question. Then there's the problem that clues come from many disparate disciplines, from biochemistry, geology, phylogenetics, ecology, chemistry and cosmology. Few can claim real expertise in all those fields. And now we are in the midst of a genomic revolution. We have thousands of complete genome sequences, codes that stretch over millions or billions of digits, all too often containing conflicting signals from the deep past. Interpreting these data demands rigorous logical, computational and statistical know-how; any biological understanding is a bonus. And so the clouds have been swirling around with arguments. Each time a gap opens up, it reveals an increasingly surreal landscape. The old comforts have been evaporating. We're now faced with a stark new picture, and it's both real and troubling. And from a researcher's standpoint, hoping to find some significant new problem to solve, it's flat out thrilling! The biggest questions in biology are yet to be solved. This book is my own attempt to make a start.

How do bacteria relate to complex life? The roots of the question date right back to the discovery of microbes by the Dutch microscopist Antony van Leeuwenhoek in the 1670s. His menagerie of 'little animals' thriving under the microscope took some believing, but was soon confirmed by the equally ingenious Robert Hooke. Leeuwenhoek also discovered bacteria, and wrote about them in a famous paper of 1677: they were 'incredibly small; nay, so small, in my sight, that I judged that even if 100 of these very wee animals lay stretched out one against another, they could not reach the length of a grain of course sand; and if this be true, then ten hundred thousand of these living creatures could scarce equal the bulk of a course grain of sand'. Many doubted that Leeuwenhoek could possibly have seen bacteria using his simple single-lens microscopes, though it is now incontrovertible that he did so. Two points stand out. He found bacteria everywhere – in rainwater and the sea, not just on his own teeth. And he intuitively made some distinction between these 'very wee animals' and the 'gygantick monsters' – microscopic protists! – with their enthralling behaviour and 'little feet' (cilia). He even noticed that some larger cells were composed of a number of little 'globules', which he compared with bacteria (though not in those terms). Among these little globules, Leeuwenhoek almost certainly saw the cell nucleus, the repository of the genes in all complex cells. And there the matter lay for several centuries. The famous classifier Carl Linnaeus, 50 years after Leeuwenhoek's discoveries, had just lumped all microbes into the genus *Chaos* (formless) of the phylum Vermes (worms). In the nineteenth century Ernst Haeckel, the great German evolutionist and contemporary of Darwin, formalised the deep distinction again, separating bacteria from other microbes. But in conceptual terms, there was little advance until the middle of the twentieth century.

The unification of biochemistry brought matters to a head. The sheer metabolic virtuosity of bacteria had made them seem uncategorisable. They can grow on anything, from concrete to battery acid to gases. If these totally different ways of life had nothing in common, how could bacteria be classified? And if not classified, how could we understand them? Just as the periodic table brought coherence to chemistry, so biochemistry brought an order to the evolution of cells. Another Dutchman, Albert Kluyver, showed that similar biochemical processes underpinned the extraordinary diversity of life. Processes as distinct as respiration, fermentation and photosynthesis all shared a common basis, a conceptual integrity which attested that all life had descended from a common ancestor. What was true of bacteria, he said, was also true of elephants. At the level of their biochemistry, the barrier between bacteria and complex cells barely exists. Bacteria are enormously more versatile, but the basic processes that keep them alive are similar. Kluyver's own student Cornelis van Niel, together with Roger Stanier, perhaps came closest to appreciating the difference: bacteria, like atoms, could not be broken down any further, they said: bacteria are the smallest unit of function. Many bacteria can respire oxygen in the same way that we do, for example, but it takes the whole bacterium to do so. Unlike our own cells, there are no internal parts dedicated to respiration. Bacteria divide in half as they grow, but in function they are indivisible.

And then came the first of three major revolutions that have wracked our view of life in the past half century. This first was instigated by Lynn Margulis in the summer of love, 1967. Complex cells did not evolve by 'standard' natural selection, Margulis argued, but in an orgy of cooperation, in which cells engaged with each other so closely that they even got inside each other. Symbiosis is a long-term interaction between two or more species, usually some sort of trade for wares or services. In the case of microbes, those wares are the substances of life, the substrates of metabolism, which power the lives of cells. Margulis talked about *endo*symbiosis – the same types of trade, but now so intimate that some collaborating cells physically live inside their host cell like the traders who sold from within the temple. These ideas trace their roots to the early twentieth century, and are reminiscent of plate tectonics. It 'looks' as if Africa and South America were once joined together, and later pulled apart, but this childlike notion was long ridiculed as absurd. Likewise, some of the structures inside complex cells look like bacteria, and even give the impression of growing and dividing independently. Perhaps the explanation really was as simple as that – they are bacteria!

Like tectonics these ideas were ahead of their time, and it was not until the era of molecular biology in the 1960s that it was possible to present a strong case. This Margulis did for two specialised structures inside cells – the mitochondria, seats of respiration, in which food is burned in oxygen to provide the energy needed for living, and the chloroplasts, the engines of photosynthesis in plants, which convert solar power into chemical energy. Both of these 'organelles' (literally miniature organs) retain tiny specialised genomes of their own, each one with a handful of genes encoding at most a few dozen proteins involved in the mechanics of respiration or photosynthesis. The exact sequences of these genes ultimately gave the game away – plainly, mitochondria and chloroplasts do derive from bacteria. But notice I say 'derive'. They are no longer bacteria, and don't have any real independence, as the vast majority of the genes needed for their existence (at least 1,500 of them) are found in the nucleus, the genetic 'control centre' of the cell.

Margulis was right about the mitochondria and chloroplasts; by the 1980s, few doubters remained. But her enterprise was much greater: for Margulis, the entire complex cell, now generally known as the *eukaryotic* cell (from the Greek meaning 'true nucleus') was a patchwork of symbioses. In her eyes, many other parts of the complex cell, notably the cilia (Leeuwenhoek's 'little feet'), also derived from bacteria (spirochetes in the case of cilia). There had been a long succession of mergers, which Margulis now formalised as the 'serial endosymbiosis theory'. Not just individual cells but the whole world was a vast collaborative network of bacteria – 'Gaia', an idea that she pioneered with James Lovelock. While the concept of Gaia has enjoyed a renaissance in the more formal guise of 'earth systems science' in recent years (stripping Lovelock's original teleology), the idea that complex 'eukaryotic' cells are an ensemble of bacteria has far less to support it. Most of the structures of the cell do not look as if they derived from bacteria, and there's nothing in the genes to suggest that they do. So Margulis was right about some things and almost certainly wrong about others. But her crusading spirit, forceful femininity, dismissal of Darwinian competition and tendency to believe in conspiracy theories, meant that when she died prematurely from a stroke in 2011, her legacy was decidedly mixed. A feminist heroine for some and loose cannon for others, much of this legacy was sadly far removed from science.

Revolution number two was the phylogenetic revolution – the ancestry of genes. The possibility had been anticipated by Francis Crick as early as 1958. With characteristic aplomb, he wrote: 'Biologists should realise that before long we shall have a subject which might be called "protein taxonomy" – the study of amino acid sequences of proteins of an organism and the comparison of them between species. It can be argued that these sequences are the most delicate expression possible of the phenotype of an organism and that vast amounts of evolutionary information may be hidden away within them.' And lo, it came to pass. Biology is now very much about the information concealed in the sequences of proteins and genes. We no longer compare the sequences of amino acids directly, but the sequences of letters in DNA (which encodes proteins), giving even greater sensitivity. Yet for all his vision, neither Crick nor anyone else began to imagine the secrets that actually emerged from the genes.

The scarred revolutionary was Carl Woese. In work beginning quietly in the 1960s and not bearing fruit until a decade later, Woese selected a single gene to compare between species. Obviously, the gene had to be present in all species. What's more, it had to serve the same purpose. That purpose had to be so fundamental, so important to the cell, that even slight changes in its function would be penalised by natural selection. If most changes are eliminated, what remains must be relatively unchanging – evolving extremely slowly, and changing little over vast periods of time. That's necessary if we want to compare the differences that accumulate between species over literally billions of years, to build a great tree of life, going back to the beginning. That was the scale of Woese's ambition. Bearing all these requirements in mind, he turned to a basic property of all cells, the ability to make proteins.

Proteins are assembled on remarkable nanomachines found in all cells, called ribosomes. Excepting the iconic double helix of DNA, nothing is more symbolic of the informational age of biology than the ribosome. Its structure also epitomises a contradiction that is hard for the human mind to fathom – scale. The ribosome is unimaginably tiny. Cells are already microscopic. We had no inkling of their existence for most of human history. Ribosomes are orders of magnitude smaller still. You have 13 *million* of them in a single cell from your liver. But ribosomes are not only incomprehensibly small; on the scale of atoms, they are massive, sophisticated superstructures. They're composed of scores of substantial subunits, moving machine parts that act with far more precision than an automated factory line. That's not an exaggeration. They draw in the 'tickertape' code-script that encodes a protein, and translate its sequence precisely, letter by letter, into the protein itself. To do so, they recruit all the building blocks (amino acids) needed, and link them together into a long chain, their order specified by the code-script. Ribosomes have an error rate of about one letter in 10,000, far lower than the defect rate in our own high-quality manufacturing processes. And they operate at a rate of about 10 amino acids per second, building whole proteins with chains comprising hundreds of amino acids in less than a minute. Woese chose one of the subunits from the ribosome, a single machine part, so to speak, and compared its sequence across different species, from bacteria such as *E. coli* to yeast to humans.

His findings were a revelation, and turned our world view on its head. He could distinguish between the bacteria and complex eukaryotes without any difficulty, laying out the branching tree of genetic relatedness within and between each of these magisterial groups. The only surprise in this was how little difference there is between plants and animals and fungi, the groups that most biologists have spent most of their lives studying. What nobody anticipated was the existence of a third domain of life. Some of these simple cells had been known for centuries, but were mistaken for bacteria. They look like bacteria. Exactly like bacteria: equally tiny, and equally lacking in discernible structure. But the difference in their ribosomes was like the smile of the Cheshire cat, betraying the presence of a different sort of absence. This new group might have lacked the complexity of eukaryotes, but the genes and proteins that they did have were shockingly different from those of bacteria. This second group of simple cells became known as the *archaea*, on the hunch that they're even older than the bacteria, which is probably not true; modern views have it that they are equally old. But at the arcane level of their genes and biochemistry, the gulf between bacteria and archaea is as great as that between bacteria and eukaryotes (us). Almost literally. In Woese's famous 'three domains' tree of life, archaea and eukaryotes are 'sister groups', sharing a relatively recent ancestor.

In some respects, the archaea and eukaryotes do indeed have a lot in common, especially in terms of information flow (the way that they read off their genes and convert them into proteins). In essence, archaea have a few sophisticated molecular machines resembling those of eukaryotes, if with fewer parts – the seeds of eukaryotic complexity. Woese refused to countenance any deep morphological gulf between bacteria and eukaryotes, but proposed three equivalent domains, each of which had explored vast realms of evolutionary space, none of which could be given precedence. Most forcefully, he rejected the old term 'prokaryote' (meaning literally 'before the nucleus', which could be applied to both archaea and bacteria) as there was nothing in his tree to suggest a genetic basis for that distinction. On the contrary, he pictured all three domains as reaching right back into the deepest past, sharing a mysterious common ancestor, from which they had somehow 'crystallised'. Towards the end of his life Woese became almost mystical about those earliest stages of evolution, calling for a more holistic view of life. That's ironic, given that the revolution he wrought was based on a wholly reductionist analysis of a single gene. There is no doubt that the bacteria, archaea and eukaryotes are genuinely distinct groups and that Woese's revolution was real; but his prescription for holism, taking whole organisms and full genomes into account, is right now ushering in the third cellular revolution – and it overturns Woese's own.

This third revolution isn't over yet. It's a little more subtle in reasoning, but packs the biggest punch of all. It is rooted in the first two revolutions, and specifically in the question: how do the two relate? Woese's tree depicts the divergence of one fundamental gene in the three domains of life. Margulis, in contrast, has genes from different species converging together in the mergers and acquisitions of endosymbiosis. Depicted as a tree, this is the fusion, not the bifurcation, of branches – the opposite of Woese. They can't both be right! Neither do they both have to be totally wrong. The truth, as so often in science, lies somewhere between the two. But don't think that makes it a compromise. The answer that's emerging is more exciting than either alternative.

We know that mitochondria and chloroplasts were indeed derived from bacteria by endosymbiosis, but that the other parts of complex cells probably evolved by conventional means. The question is: when, exactly? Chloroplasts are found only in algae and plants, hence were most likely acquired in an ancestor of those groups alone. That puts them as a relatively late acquisition. Mitochondria, in contrast, are found in all eukaryotes (there's a backstory there that we'll examine in Chapter 1) and so must have been an earlier acquisition. But how early? Put another way, what kind of cell picked up mitochondria? The standard textbook view is that it was quite a sophisticated cell, something like an amoeba, a predator that could crawl around, change shape and engulf other cells by a process known as phagocytosis. In other words, mitochondria were acquired by a cell that was not so far from being a fully fledged, card-carrying eukaryote. We now know that's wrong. Over the last few years, comparisons of large numbers of genes in more representative samples of species have come to the unequivocal conclusion that the host cell was in fact an archaeon – a cell from the domain Archaea. All archaea are prokaryotes. By definition, they don't have a nucleus or sex or any of the other traits of complex life, including phagocytosis. In terms of its morphological complexity, the host cell must have had next to nothing. Then, somehow, it acquired the bacteria that went on to become mitochondria. Only *then* did it evolve all those complex traits. If so, the singular origin of complex life might have *depended* on the acquisition of mitochondria. They somehow triggered it.

This radical proposition – complex life arose from a singular endosymbiosis between an archaeon host cell and the bacteria that became mitochondria – was predicted by the brilliantly intuitive and free-thinking evolutionary biologist Bill Martin, in 1998, on the basis of the extraordinary mosaic of genes in eukaryotic cells, a mosaic largely uncovered by Martin himself. Take a single biochemical pathway, say fermentation. Archaea do it one way, and bacteria quite a different way; the genes involved are distinct. Eukaryotes have taken a few genes from bacteria, and a few others from archaea, and woven them together into a tightly knit composite pathway. This intricate fusion of genes doesn't merely apply to fermentation, but to almost all biochemical processes in complex cells. It is an outrageous state of affairs!

Martin thought all this through in great detail. Why did the host cell pick up so many genes from its own endosymbionts, and why did it integrate them so tightly into its own fabric, replacing many of its existing genes in the process? His answer, with Miklós Müller, is called the hydrogen hypothesis. Martin and Müller argued that the host cell was an archaeon, capable of growing from two simple gases, hydrogen and carbon dioxide. The endosymbiont (the future mitochondrion) was a versatile bacterium (perfectly normal for bacteria), which provided its host cell with the hydrogen it needed to grow. The details of this relationship, worked out step by step on a logical basis, explain why a cell that started out living from simple gases would end up scavenging organics (food) to supply its own endosymbionts. But that's not the important point for us here. The salient point is: Martin predicted that complex life arose through a *singular* endosymbiosis between two cells only. He predicted that the host cell was an archaeon, lacking the baroque complexity of eukaryotic cells. He predicted that there never was an intermediate, simple eukaryotic cell, which lacked mitochondria; the acquisition of mitochondria and the origin of complex life was one and the same event. And he predicted that all the elaborate traits of complex cells, from the nucleus to sex to phagocytosis, evolved *after* the acquisition of mitochondria, in the context of that unique endosymbiosis. This is one of the finest insights in evolutionary biology, and deserves to be much better known. It would be, were it not so easily confounded with the serial endosymbiosis theory (which we'll see makes none of the same predictions). All of these explicit predictions have been borne out in full by genomic research over the past two decades. It's a monument to the power of biochemical logic. If there were a Nobel Prize in Biology, nobody would be a more deserving recipient than Bill Martin.

And so we have come full circle. We know an awful lot, but we still don't know why life is the way it is. We know that complex cells arose on just one occasion in 4 billion years of evolution, through a singular endosymbiosis between an archaeon and a bacterium (**Figure 1**). We know that the traits of complex life arose in the aftermath of this union; but we still do not know why those particular traits arose in eukaryotes, while showing no signs of evolving in either bacteria or archaea. We don't know what forces constrain bacteria and archaea – why they remain morphologically simple, despite being so different in their biochemistry, so varied in their genes, so versatile in their ability to extract a living from gases and rocks. What we do have is a radical new framework in which to approach the problem.

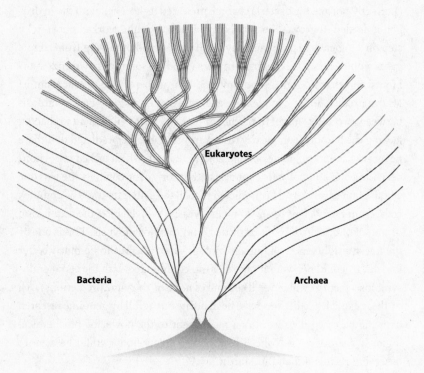

Figure 1 **A tree of life showing the chimeric origin of complex cells**
A composite tree reflecting whole genomes, as depicted by Bill Martin in 1998, showing the three domains of Bacteria, Archaea and Eukaryotes. Eukaryotes have a chimeric origin, in which genes from an archaeal host cell and a bacterial endosymbiont coalesce, with the archaeal host cell ultimately evolving into the morphologically complex eukaryotic cell, and the endosymbionts into mitochondria. One group of eukaryotes later acquired a second bacterial endosymbiont, which became the chloroplasts of algae and plants.

I believe the clue lies in the bizarre mechanism of biological energy generation in cells. This strange mechanism exerts pervasive but little appreciated physical constraints on cells. Essentially all living cells power themselves through the flow of protons (positively charged hydrogen atoms), in what amounts to a kind of electricity – proticity – with protons in place of electrons. The energy we gain from burning food in respiration is used to pump protons across a membrane, forming a reservoir on one side of the membrane. The flow of protons back from this reservoir can be used to power work in the same way as a turbine in a hydroelectric dam. The use of cross-membrane proton gradients to power cells was utterly unanticipated. First proposed in 1961 and developed over the ensuing three decades by one of the most original scientists of the twentieth century, Peter Mitchell, this conception has been called the most counterintuitive idea in biology since Darwin, and the only one that compares with the ideas of Einstein, Heisenberg and Schrödinger in physics. At the level of proteins, we now know how proton power works in detail. We also know that the use of proton gradients is universal across life on earth – proton power is as much an integral part of all life as the universal genetic code. Yet we know next to nothing about how or why this counterintuitive mechanism of energy harnessing first evolved. So it seems to me there are two big unknowns at the very heart of biology today: why life evolved in the perplexing way it did, and why cells are powered in such a peculiar fashion.

This book is an attempt to answer these questions, which I believe are tightly entwined. I hope to persuade you that energy is central to evolution, that we can only understand the properties of life if we bring energy into the equation. I want to show you that this relationship between energy and life goes right back to the beginning – that the fundamental properties of life necessarily emerged from the disequilibrium of a restless planet. I want to show you that the origin of life was driven by energy flux, that proton gradients were central to the emergence of cells, and that their use constrained the structure of both bacteria and archaea. I want to demonstrate that these constraints dominated the later evolution of cells, keeping the bacteria and archaea forever simple in morphology, despite their biochemical virtuosity. I want to prove that a rare event, an endosymbiosis in which one bacterium got inside an archaeon, broke those constraints, enabling the evolution of vastly more complex cells. I want to show you that this was not easy – that the intimate relationship between cells living one inside another explains why morphologically complex organisms only arose once. I hope to do more, to persuade you that this intimate relationship actually predicts some of the properties of complex cells. These traits include the nucleus, sex, two sexes, and even the distinction between the immortal germline and the mortal body – the origins of a finite lifespan and genetically predetermined death. Finally, I want to convince you that thinking in these energetic terms allows us to predict aspects of our own biology, notably a deep evolutionary trade-off between fertility and fitness in youth, on the one hand, and ageing and disease on the other. I'd like to think that these insights might help us to improve our own health, or at least to understand it better.

It can be frowned upon to act as an advocate in science, but there is a fine tradition of doing exactly that in biology, going back to Darwin himself; he called *The Origin of Species* 'one long argument'. A book is still the best way to lay out a vision of how facts might relate to each other across the whole fabric of science – a hypothesis that makes sense of the shape of things. Peter Medawar described a hypothesis as an imaginative leap into the unknown. Once the leap is taken, a hypothesis becomes an attempt to tell a story that is understandable in human terms. To be science, the hypothesis must make predictions that are testable. There's no greater insult in science than to say that an argument is 'not even wrong', that it is invulnerable to disproof. In this book, then, I will lay out a hypothesis – tell a coherent story – that connects energy and evolution. I will do so in enough detail that I can be proved wrong, while writing as accessibly and as excitingly as I can. This story is based in part on my own research (you'll find the original papers in Further Reading) and in part on that of others. I have collaborated most fruitfully with Bill Martin in Düsseldorf, who I've found has an uncanny knack of being right, and Andrew Pomiankowski, a mathematically minded evolutionary geneticist and best of colleagues at University College London; and with several extremely able PhD students. It has been a privilege and an enormous pleasure, and we are only at the beginning of an immense journey.

I have tried to keep this book short and to the point, to cut down on digressions and interesting but unrelated stories. The book is an argument, as spare or detailed as it needs to be. It is not lacking in metaphor and (I hope) entertaining details; that is vital to bring a book grounded in biochemistry to life for the general reader. Few of us can easily visualise the alien submicroscopic landscape of giant interacting molecules, the very stuff of life. But the point is the science itself, and that has fashioned my writing. To call a spade a spade is a good old-fashioned virtue. It's succinct, and brings us straight to the point; you would soon become irritated if I insisted on reminding you every few pages that a spade is a digging implement used for burying people. While it's less helpful to call a mitochondrion a mitochondrion, it's likewise cumbersome to keep writing "All large, complex cells such as our own contain miniature powerhouses, which derived long ago from free-living bacteria, and which today provide essentially all our energy needs." I could write instead: "All eukaryotes have mitochondria." This is clearer and packs a greater punch. When you are comfortable with a few terms, they convey more information, and so succinctly that in this case they immediately beg a question: how did that come about? That leads straight to the edge of the unknown, to the most interesting science. So I've tried to avoid unnecessary jargon, and have included occasional reminders of the meaning of terms; but beyond that I hope you will gain familiarity with recurring terms. As a failsafe, I've also included a short glossary of the main terms at the end. With the occasional double check, I hope this book will be wholly accessible to anyone who is interested.

And I sincerely hope you will be interested! For all its strangeness, this brave new world is genuinely exciting: the ideas, the possibilities, the dawning understanding of our place in this vast universe. I will outline the contours of a new and largely uncharted landscape, a perspective that stretches from the very origin of life to our own health and mortality. This colossal span is thankfully united by a few simple ideas that relate to proton gradients across membranes. For me the best books in biology, ever since Darwin, have been arguments. This book aspires to follow in that tradition. I will argue that energy has constrained the evolution of life on earth; that the same forces ought to apply elsewhere in the universe; and that a synthesis of energy and evolution could be the basis for a more predictive biology, helping us understand why life is the way it is, not only on earth, but wherever it might exist in the universe.

PART I

THE PROBLEM

WHAT IS LIFE?

U nblinking day and night, the radio telescopes scrutinise the skies. Forty-two of them are scattered in a loose cluster across the scrubby sierra of northern California. Their white bowls resemble blank faces, all focused hopefully in unison on some point beyond the horizon, as if this were an assembly point for alien invaders trying to go home. The incongruity is apt. The telescopes belong to SETI, the search for extraterrestrial intelligence, an organisation that has been scanning the heavens for signs of life for half a century, to no avail. Even the protagonists are not too optimistic about its chances of success; but when funding dried up a few years ago, a direct appeal to the public soon had the Allen Telescope Array operational again. To my mind the venture is a poignant symbol of humanity's uncertain sense of our place in the universe, and indeed the fragility of science itself: science fiction technology so inscrutable that it hints at omniscience, trained on a dream so naive that it is barely grounded in science at all, that we are not alone.

Even if the array never detects life, it is still valuable. It may not be possible to look the wrong way through these telescopes, yet that is their real power. What exactly are we looking for out there? Should life elsewhere in the universe be so similar to us that it too uses radio waves? Do we think that life elsewhere should be carbon based? Would it need water? Oxygen? These are not really questions about the make-up of life somewhere else in the universe: they are about life on earth, about why life is the way we know it. These telescopes are mirrors, reflecting their questions back at earthly biologists. The problem is that science is all about predictions. The most pressing questions in physics are about *why* the laws of physics are as they are: what fundamental principles predict the known properties of the universe? Biology is less predictive, and has no laws to compare with those of physics, but even so the predictive power of evolutionary biology is embarrassingly bad. We know a great deal about the molecular mechanisms of evolution and about the history of life on our planet, but far less about which parts of this history are chance – trajectories that could have played out quite differently on other planets – and which bits are dictated by physical laws or constraints.

That is not for any lack of effort. This terrain is the playground of retired Nobel laureates and other towering figures in biology; yet for all their learning and intellect, they cannot begin to agree among themselves. Forty years ago, at the dawn of molecular biology, the French biologist Jacques Monod wrote his famous book *Chance and Necessity*, which argues bleakly that the origin of life on earth was a freak accident, and that we are alone in an empty universe. The final lines of his book are close to poetry, an amalgam of science and metaphysics:

> The ancient covenant is in pieces; man knows at last that he is alone in the universe's unfeeling immensity, out of which he emerged only by chance. His destiny is nowhere spelled out, nor is his duty. The kingdom above or the darkness below: it is for him to choose.

Since then, others have argued the opposite: that life is an inevitable outcome of cosmic chemistry. It will arise quickly, almost everywhere. Once life is thriving on a planet, what happens next? Again, there is no consensus. Engineering constraints may force life down convergent pathways to similar places, regardless of where it starts. Given gravity, animals that fly are likely to be lightweight, and possess something akin to wings. In a more general sense, it may be necessary for life to be cellular, composed of small units that keep their insides different from the outside world. If such constraints are dominant, life elsewhere may closely resemble life on earth. Conversely, perhaps contingency rules – the make-up of life depends on the random survivors of global accidents such as the asteroid impact that wiped out the dinosaurs. Wind back the clock to Cambrian times, half a billion years ago, when animals first exploded into the fossil record, and let it play forwards again. Would that parallel world be similar to our own? Perhaps the hills would be crawling with giant terrestrial octopuses.

One of the reasons for pointing telescopes out to space is that here on earth we are dealing with a sample size of one. From a statistical point of view, we can't say what, if anything, constrained the evolution of life on earth. But if that were really true, there would be no basis for this book, or any others. The laws of physics apply throughout the universe, as do the properties and abundance of elements, hence the plausible chemistry. Life on earth has many strange properties that have taxed the minds of the finest biologists for centuries – traits like sex and ageing. If we could predict from first principles – from the chemical make-up of the universe – why such traits arose, why life is the way it is, then we would have access to the world of statistical probability again. Life on earth is not really a sample of one, but for practical purposes it is an infinite variety of organisms evolving over infinite time. Yet evolutionary theory does not predict, from first principles, why life on earth took the course that it did. I do not mean by this that I think evolutionary theory is wrong – it is not – but simply that it is not predictive. My argument in this book is that there are in fact strong constraints on evolution – energetic constraints – which do make it possible to predict some of the most fundamental traits of life from first principles. Before we can address these constraints, we must consider why evolutionary biology is not predictive, and why these energetic constraints have passed largely unnoticed; indeed, why we have hardly even noticed there is a problem. It has only become starkly apparent in the past few years, and only to those who follow evolutionary biology, that there is a deep and disturbing discontinuity at the very heart of biology.

To a point, we can blame DNA for this sorry state of affairs. Ironically, the modern era of molecular biology, and all the extraordinary DNA technology that it entails, arguably began with a physicist, specifically with the publication of Erwin Schrödinger's book *What is Life?* in 1944. Schrödinger made two key points: first, that life somehow resists the universal tendency to decay, the increase in entropy (disorder) that is stipulated by the second law of thermodynamics; and second, that the trick to life's local evasion of entropy lies in the genes. He proposed that the genetic material is an 'aperiodic' crystal, which does not have a strictly repeating structure, hence could act as a 'code-script' – reputedly the first use of the term in the biological literature. Schrödinger himself assumed, along with most biologists at the time, that the quasicrystal in question must be a protein; but within a frenzied decade, Crick and Watson had inferred the crystal structure of DNA itself. In their second *Nature* paper of 1953, they wrote: 'It therefore seems likely that the precise sequence of the bases is the code which carries the genetical information.' That sentence is the basis of modern biology. Today biology is information, genome sequences are laid out *in silico*, and life is defined in terms of information transfer.

Genomes are the gateway to an enchanted land. The reams of code, 3 billion letters in our own case, read like an experimental novel, an occasionally coherent story in short chapters broken up by blocks of repetitive text, verses, blank pages, streams of consciousness: and peculiar punctuation. A tiny proportion of our own genome, less than 2%, codes for proteins; a larger portion is regulatory; and the function of the rest is liable to cause intemperate rows among otherwise polite scientists.[1] It doesn't matter here. What is clear is that genomes can encode up to tens of thousands of genes and a great deal of regulatory complexity, capable of specifying everything that is needed to transform a caterpillar into a butterfly or a child into an adult human. Comparing the genomes of animals, plants, fungi, and

---

1 There is a noisy dispute about whether all this non-coding DNA serves any useful purpose. Some claim that it does, and that the term 'junk DNA' should be dismissed. Others pose the 'onion test': if most non-coding DNA does serve a useful purpose, why does an onion need five times more of it than a human being? In my view it's premature to abandon the term. Junk is not the same thing as garbage. Garbage is thrown away immediately; junk is retained in the garage, in the hope that it might turn out to be useful one day.

single-celled amoebae shows that the same processes are at play. We can find variants of the same genes, the same regulatory elements, the same selfish replicators (such as viruses) and the same stretches of repetitive nonsense in genomes of vastly different sizes and types. Onions, wheat and amoebae have more genes and more DNA than we do. Amphibians such as frogs and salamanders have genome sizes that range over two orders of magnitude, with some salamander genomes being 40 times larger than our own, and some frogs being less than a third of our size. If we had to sum up the architectural constraints on genomes in a single phrase, it would have to be 'anything goes'.

That's important. If genomes are information, and there are no fundamental constraints on genome size and structure, then there are no constraints on information either. That doesn't mean there are no constraints on genomes at all. Obviously there are. The forces acting on genomes include natural selection as well as more random factors – accidental duplications of genes, chromosomes or whole genomes, inversions, deletions, and invasions of parasitic DNA. How all of this pans out depends on factors such as niche, competition between species, and population size. From our point of view, all these factors are unpredictable. They are part of the environment. If the environment is specified precisely, we might just be able to predict the genome size of a particular species. But an infinite number of species live in an endless variety of microenvironments, ranging from the insides of other cells, to human cities, to pressurised ocean depths. Not so much 'anything goes' as 'everything goes'. We should expect to find as much variety in genomes as there are factors acting on them in these diverse environments. Genomes do not predict the future but recall the past: they reflect the exigencies of history.

Consider other worlds again. If life is about information, and information is unconstrained, then we can't predict what life might look like on another planet, only that it will not contravene the laws of physics. As soon as some form of hereditary material has arisen – whether DNA or something else – then the trajectory of evolution becomes unconstrained by information and unpredictable from first principles. What actually evolves will depend on the exact environment, the contingencies of history, and the ingenuity of selection. But look back to earth. This statement is reasonable for the enormous variety of life as it exists today; but it is simply not true for most of the long history of the earth. For billions of years, it seems that life was constrained in ways that can't easily be interpreted in terms of genomes, history or environment. Until recently, the peculiar history of life on our planet was far from clear, and even now there is much ado about the details. Let me sketch out the emerging view, and contrast it with older versions that now look to be wrong.

A brief history of the first 2 billion years of life

Our planet is about 4.5 billion years old (that's to say 4,500 million years old). It was wracked during its early history, for some 700 million years, by a heavy asteroid bombardment, as the nascent solar system settled itself down. A colossal early impact with a Mars-sized object probably formed the moon. Unlike the earth, whose active geology continuously turns over the crust, the pristine surface of the moon preserves evidence for this early bombardment in its craters, dated by rocks brought home by the Apollo astronauts.

Despite the absence of earthly rocks of a comparable age, there are still a few clues to conditions on the early earth. In particular, the composition of zircons (tiny crystals of zirconium silicate, smaller than sand grains, found in many rocks) suggests that there were oceans much earlier than we had thought. We can tell from uranium dating that some of these amazingly robust crystals were formed between 4 and 4.4 billion years ago, and later accumulated as detrital grains in sedimentary rocks. Zircon crystals behave like tiny cages that trap chemical contaminants, reflecting the environment in which they were formed. The chemistry of early zircons suggests that they were formed at relatively low temperatures, and in the presence of water. Far from the image of a volcanic hell, with oceans of boiling lava captured vividly in artists' impressions of the what is technically termed the 'Hadean' period, zircon crystals point to a more tranquil water world with a limited land surface.

Likewise, the old idea of a primordial atmosphere replete with gases such as methane, hydrogen and ammonia, which react together to form organic molecules, does not stand the scrutiny of zircons. Trace elements such as cerium are incorporated into zircon crystals mostly in their oxidised form. The high content of cerium in the earliest zircons suggests that the atmosphere was dominated by oxidised gases emanating from volcanoes, notably carbon dioxide, water vapour, nitrogen gas and sulphur dioxide. This mixture is not dissimilar in composition to the air today, except that it was missing oxygen itself, which was not abundant until much later, after the advent of photosynthesis. Reading the make-up of a long-vanished world from a few scattered zircon crystals puts a lot of weight on what amounts to grains of sand, but it is better than no evidence at all. That evidence consistently conjures up a planet that was surprisingly similar to the one we know today. The occasional asteroid impact might have partially vaporised the oceans, but is unlikely to have upset any bacteria living in the deep oceans – if they had already evolved.

The earliest evidence for life is equally flimsy, but may date back to some of the earliest known rocks at Isua and Akilia in south-west Greenland, which are around 3.8 billion years old (see **Figure 2** for a timeline). This evidence is not in the form of fossils or complex molecules derived from living cells ('biomarkers') but is simply a non-random sorting of carbon atoms in graphite. Carbon comes in two stable forms, or isotopes, which have marginally different masses.[2] Enzymes (proteins that catalyse reactions in living cells) have a slight preference for the lighter form, carbon-12, which therefore tends to accumulate in organic matter. You could think of carbon atoms as tiny ping-pong balls – the slightly smaller balls bounce around slightly faster, so are more likely to bump into enzymes, so are more likely to be converted into organic carbon. Conversely, the heavier form, carbon-13, which constitutes just 1.1% of the total carbon, is more likely to be left behind in the oceans and can instead accumulate when carbonate is precipitated out in sedimentary rocks such as limestone. These tiny

---

2 There's also a third, unstable isotope, carbon-14, which is radioactive, breaking down with a half-life of 5,570 years. This is often used for dating human artefacts, but is no use over geological periods, and so not relevant to our story here.

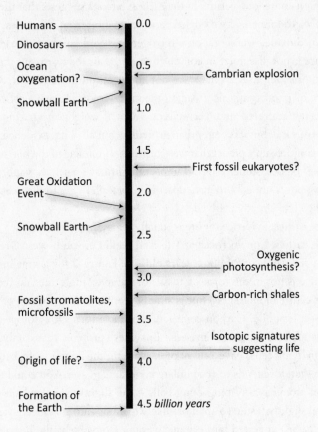

Humans → 0.0
Dinosaurs →
0.5
Ocean oxygenation? → ← Cambrian explosion
Snowball Earth → 1.0

1.5
← First fossil eukaryotes?
Great Oxidation Event → 2.0
Snowball Earth → 2.5
Oxygenic photosynthesis? ←
3.0
Carbon-rich shales ←
Fossil stromatolites, microfossils → 3.5
Isotopic signatures suggesting life ←
Origin of life? → 4.0
Formation of the Earth → 4.5 *billion years*

Figure 2 **A timeline of life**
The timeline shows approximate dates for some key events in early evolution.
Many of these dates are uncertain and disputed, but most evidence suggests that
the bacteria and archaea arose around 1.5 to 2 billion years before the eukaryotes.

differences are consistent to the point that they are often seen as diagnostic of life. Not only carbon but other elements such as iron, sulphur and nitrogen are also fractionated by living cells in a similar way. Such isotopic fractionation is reported in the graphite inclusions at Isua and Akilia.

Every aspect of this work, from the age of the rocks themselves to the very existence of the small carbon grains purported to signify life, has been challenged. What's more, it has become clear that isotopic fractionation is not unique to life at all, but can be mimicked, if more weakly, by geological processes in hydrothermal vents. If the Greenland rocks really are as ancient as they seem, and do indeed contain fractionated carbon, that is still no proof of life. This might seem discouraging, but in another sense is no less than we should expect. I shall argue that the distinction between a 'living planet' – one that is geologically active – and a living cell is only a matter of definition. There is no hard and fast dividing line. Geochemistry gives rise seamlessly to biochemistry. From this point of view, the fact that we can't distinguish between geology and biology in these old rocks is fitting. Here is a living planet giving rise to life, and the two can't be separated without splitting a continuum.

Move forward a few hundred million years and the evidence for life is more tangible – as solid and scrutable as the ancient rocks of Australia and South Africa. Here, there are microfossils that look a lot like cells, although trying to place them in modern groups is a thankless task. Many of these tiny fossils are lined with carbon, again featuring telltale isotopic signatures, but now somewhat more consistent and pronounced, suggesting organised metabolism rather than haphazard hydrothermal processes. And there are structures resembling stromatolites, those domed cathedrals of bacterial life, in which cells grow layer upon layer, the buried layers mineralising, turning to stone, ultimately building up into strikingly laminated rock structures, a metre in height. Beyond these direct fossils, by 3.2 billion years ago there are large-scale geological features, hundreds of square miles in area and tens of metres deep, notably banded iron formations and carbon-rich shales. We tend to think of bacteria and minerals as occupying different realms, living versus inanimate, but in fact many sedimentary rocks are deposited, on a colossal scale, by bacterial processes. In the case of banded-iron formations – stunningly beautiful in their stripes of red and black – bacteria strip electrons from iron dissolved in the oceans (such 'ferrous' iron is plentiful in the absence of oxygen) leaving behind the insoluble carcass, rust, to sink down to the depths. Why these iron-rich rocks are striped remains puzzling, but isotope signatures again betray the hand of biology.

These vast deposits indicate not just life but photosynthesis. Not the familiar form of photosynthesis that we see around us in the green leaves of plants and algae, but a simpler precursor. In all forms of photosynthesis, the energy of light is used to strip electrons from an unwilling donor. The electrons are then forced on to carbon dioxide to form organic molecules. The various forms of photosynthesis differ in their source of electrons, which can come from all kinds of different places, most commonly dissolved (ferrous) iron, hydrogen sulphide, or water. In each case, electrons are transferred to carbon dioxide, leaving behind the waste: rusty iron deposits, elemental sulphur (brimstone) and oxygen, respectively. The hardest nut to crack, by far, is water. By 3.2 billion years ago, life was extracting electrons from almost everything else. Life, as biochemist Albert Szent-Györgyi observed, is nothing but an electron looking for a place to rest. Quite when the final step to extracting electrons from water took place is contentious. Some claim it was an early event in evolution, but the weight of evidence now suggests that 'oxygenic' photosynthesis arose between 2.9 and 2.4 billion years ago, not so long before a cataclysmic period of global unrest, the earth's midlife crisis. Worldwide glaciations, known as a 'snowball earth', were followed by the widespread oxidation of terrestrial rocks, around 2.2 billion years ago, leaving rusty 'red beds' as a definitive sign of oxygen in the air – the 'Great Oxidation Event'. Even the global glaciations indicate a rise of atmospheric oxygen. By oxidising methane, oxygen removed a potent greenhouse gas from the air, triggering the global freeze.[3]

---

3 This methane was produced by methanogenic bacteria, or more specifically archaea, which if carbon isotope signatures are to be believed (methanogens produce a particularly strong signal), were thriving before 3.4 billion years ago. As noted earlier, methane was not a significant constituent of the earth's primordial atmosphere.

With the evolution of oxygenic photosynthesis, life's metabolic tool kit was essentially complete. Our whistle-stop tour through nearly 2 billion years of earth's history – three times longer than the entire duration of animals – is unlikely to be accurate in all its details, but it is worth pausing a moment to consider what the bigger picture says about our world. First, life arose very early, probably between 3.5 and 4 billion years ago, if not earlier, on a water world not unlike our own. Second, by 3.5 to 3.2 billion years ago, bacteria had already invented most forms of metabolism, including multiple forms of respiration and photosynthesis. For a billion years the world was a cauldron of bacteria, displaying an inventiveness of biochemistry that we can only wonder at.[4] Isotopic fractionation suggests that all the major nutrient cycles – carbon, nitrogen, sulphur, iron, and so on – were in place before 2.5 billion years ago. Yet only with the rise of oxygen, from 2.4 billion years ago, did life transfigure our planet to the point that this thriving bacterial world could have been detected as a living planet from space. Only then did the atmosphere begin to accumulate a reactive mixture of gases, such as oxygen and methane, which are replenished continuously by living cells, betraying the hand of biology on a planetary scale.

The problem with genes and environment

The Great Oxidation Event has long been recognised as a pivotal moment in the history of our living planet, but its significance has shifted radically in recent years, and the new interpretation is critical to my argument in this book. The old version sees oxygen as the critical *environmental* determinant of life. Oxygen does not specify what will evolve, the argument goes, but it permits the evolution of far greater complexity – it releases the brakes. Animals, for example, make their living by physically moving around, chasing prey or being chased themselves. Obviously this requires a lot of energy, so it's easy to imagine that animals could not exist in the absence of

---

4 For most of this chapter I will refer only to bacteria for simplicity, although I mean prokaryotes, including both bacteria and archaea, as discussed in the Introduction. We'll return to the significance of archaea towards the end of the chapter.

oxygen, which provides nearly an order of magnitude more energy than other forms of respiration.[5] This statement is so blandly uninteresting it is hardly worth challenging. That is part of the problem: it does not invite further consideration. We can take it for granted that animals need oxygen (even though that is not always true), and have it, so oxygen is a common denominator. The real problems in evolutionary biology are then about the properties and behaviour of animals or plants. Or so it would seem.

This view implicitly underpins the textbook history of the earth. We tend to think of oxygen as wholesome and good, but in fact from the point of view of primordial biochemistry it is anything but: it is toxic and reactive. As oxygen levels rose, the textbook story goes, this dangerous gas put a heavy selection pressure on the whole microbial world. There are stark tales of the mass extinction to end them all – what Lynn Margulis termed the oxygen 'holocaust'. The fact that there is no trace of this cataclysm in the fossil record need not worry us too much (we are assured): these bugs were very small and it was all a terribly long time ago. Oxygen forced new relationships between cells – symbioses and endosymbioses, in which cells traded among and within themselves for the tools of survival. Over hundreds of millions of years, complexity gradually increased, as cells learned not only to deal with oxygen, but also to profit from its reactivity: they evolved aerobic respiration, giving them far more power. These large, complex, aerobic cells package their DNA in a specialised compartment called the nucleus, bequeathing their name 'eukaryotes' – literally, 'true nucleus'. I reiterate, this is a textbook story: I shall argue that it is wrong.

Today, all the complex life we see around us – all plants, animals, algae, fungi and protists (large cells like amoebae) – are composed of these

---

5 This is not strictly true. Aerobic respiration does produce nearly an order of magnitude more usable energy than fermentation, but fermentation is not technically a form of respiration at all. True anaerobic respiration uses substances other than oxygen, such as nitrate, as an electron acceptor, and these provide nearly as much energy as oxygen itself. But these oxidants can only accumulate at levels suitable for respiration in an aerobic world, as their formation depends on oxygen. So even if aquatic animals could respire using nitrate instead of oxygen, they could still only do so in an oxygenated world.

eukaryotic cells. The eukaryotes rose steadily to dominance over a billion years, the story continues, in a period known, ironically, as the 'boring billion', as little happened in the fossil record. Still, between 1.6 and 1.2 billion years ago we do begin to find fossils of single cells that look a lot like eukaryotes, some of which even fit snugly into modern groups such as red algae and fungi.

Then came another period of global unrest and a succession of snowball earths around 750–600 million years ago. Soon after that, oxygen levels rose rapidly to nearly modern levels – and the first fossils of animals appear abruptly in the fossil record. The earliest large fossils – up to a metre in diameter – are a mysterious group of symmetrical frond-like forms that most palaeontologists interpret as filter-feeding animals, though some insist are merely lichens: the Ediacarans, or more affectionately, vendobionts. Then, as abruptly as they appeared, most of these forms vanished in a mass extinction of their own, to be replaced at the dawn of the Cambrian era, 541 million years ago – a date as iconic among biologists as 1066 or 1492 – by an explosion of more recognisable animals. Large and motile, with complex eyes and alarming appendages, these fierce predators and their fearsome armour-plated prey burst on to the evolutionary scene, red in tooth and claw, Darwin in modern guise.

How much of this scenario is in fact wrong? At face value it seems to be plausible. But in my view the subtext is wrong; and as we learn more, so are a good many of the details. The subtext relates to the interplay between genes and environment. The entire scenario revolves around oxygen, supposedly the key environmental variable, which permitted genetic change, releasing the brakes on innovation. Oxygen levels rose twice, in the Great Oxidation Event 2.4 billion years ago and again towards the end of the eternal Precambrian period, 600 million years ago (**Figure 2**). Each time, the story goes, rising oxygen released constraints on structure and function. After the Great Oxidation Event, with its new threats and opportunities, cells traded among themselves in a series of endosymbioses, gradually accruing the complexity of true eukaryotic cells. When oxygen levels rose a second time, before the Cambrian explosion, the physical constraints were swept aside completely, as if with a flourish of a magician's cloak, revealing the possibility of animals for the first time. Nobody claims that oxygen physically drove these changes; rather, it transformed the selective landscape. Across the magnificent vistas of this unconstrained new landscape, genomes expanded freely, their information content finally unfettered. Life flourished, filling all conceivable niches, in every which way.

This view of evolution can be seen in terms of dialectical materialism, true to the principles of some leading evolutionary biologists during the neo-Darwinian synthesis of the early to mid twentieth century. The interpenetrating opposites are genes and environment, otherwise known as nature and nurture. Biology is all about genes, and their behaviour is all about the environment. What else is there, after all? Well, biology is not only about genes and environment, but also cells and the constraints of their physical structure, which we shall see have little to do with either genes or environment directly. The predictions that arise from these disparate world views are strikingly different.

Take the first possibility, interpreting evolution in terms of genes and the environment. The lack of oxygen on the early earth is a major environmental constraint. Add oxygen and evolution flourishes. All life that is exposed to oxygen is affected in one way or another and must adapt. Some cells just happen to be better suited to aerobic conditions, and proliferate; others die. But there are many different microenvironments. Rising oxygen doesn't simply flood the whole world with oxygen in a kind of monomaniacal global ecosystem, but it oxidises minerals on land and solutes in the oceans, and that enriches anaerobic niches too. The availability of nitrate, nitrite, sulphate, sulphite, and so on, rises. These can all be used instead of oxygen in cell respiration, so anaerobic respiration flourishes in an aerobic world. All of that adds up to many different ways of making a living in the new world.

Imagine a random mixture of cells in an environment. Some cells such as amoebae make their living by physically engulfing other cells, a process called phagocytosis. Some are photosynthetic. Others, such as fungi, digest their food externally – osmotrophy. Assuming that cell structure doesn't impose insuperable constraints, we would predict that these different cell types would descend from various different bacterial ancestors. One ancestral cell happened to be a bit better at some primitive form of phagocytosis, another at a simple form of osmotrophy, another at photosynthesis. Over time their descendants became more specialised and better adapted to that particular mode of life.

To put that more formally, if rising oxygen levels permitted flourishing new life styles, we would expect to see a *polyphyletic radiation*, in which unrelated cells or organisms (from different phyla) adapt swiftly, radiating new species that fill unoccupied niches. This kind of pattern is exactly what we do see – sometimes. Dozens of different animal phyla radiated in the Cambrian explosion, for example, from sponges and echinoderms to arthropods and worms. These great animal radiations were accompanied by matching radiations among algae and fungi, as well as protists such as ciliates. Ecology became enormously more complex, and this in itself drove further changes. Whether or not it was specifically the rising tide of oxygen that triggered the Cambrian explosion, there is general agreement that environmental changes did indeed transform selection. Something happened, and the world changed forever.

Contrast this pattern with what we would expect to see if constraints of structure dominated. Until the constraint is overcome, we should see limited change in response to any environmental shifts. We would expect long periods of stasis, impervious to environmental changes, with very occasional *monophyletic* radiations. That's to say that if, on a rare occasion, one particular group overcomes its intrinsic structural constraints, it alone will radiate to fill vacant niches (albeit possibly delayed until permitted by a change in the environment). Of course, we see this too. In the Cambrian explosion we see the radiation of different animal groups – but not multiple origins of animals. All animal groups share a common ancestor, as indeed do all plants. Complex multicellular development, involving a distinct germline and soma (body) is difficult. The constraints here relate in part to the requirements for a precise developmental program, which exercises tight control over the fate of individual cells. At a looser level, though, some degree of multicellular development is common, with as many as thirty separate origins of multicellularity among groups including algae (seaweeds), fungi and slime moulds. But there is one place in which it seems that the constraints of physical structure – cell structure – dominate to such a degree that they overwhelm everything else: the origin of the eukaryotic cell (large complex cells) from bacteria, in the aftermath of the Great Oxidation Event.

The black hole at the heart of biology

If complex eukaryotic cells really did evolve in response to the rise in atmospheric oxygen, we would predict a *polyphyletic* radiation, with various different groups of bacteria begetting more complex cell types independently. We would expect to see photosynthetic bacteria giving rise to larger and more complex algae, osmotrophic bacteria to fungi, motile predatory cells to phagocytes, and so on. Such evolution of greater complexity could occur through standard genetic mutations, gene swapping and natural selection, or by way of the mergers and acquisitions of endosymbiosis, as conceived by Lynn Margulis in the well-known serial endosymbiosis theory. Either way, if there are no fundamental constraints on cell structure, then rising oxygen levels should have made greater complexity possible regardless of how exactly it evolved. We would predict that oxygen would release the constraints on all cells, enabling a polyphyletic radiation with all kinds of different bacteria becoming more complex independently. But that's not what we see.

Let me spell this out in more detail, as the reasoning is critical. If complex cells arose via 'standard' natural selection, in which genetic mutations give rise to variations acted upon by natural selection, then we would expect to see a mixed bag of internal structures, as varied as the external appearance of cells. Eukaryotic cells are wonderfully varied in their size and shape, from giant leaf-like algal cells to spindly neurons, to outstretched amoebae. If eukaryotes had evolved most of their complexity in the course of adapting to distinct ways of life in divergent populations, then this long history should be reflected in their distinctive internal structures too. But look inside (as we'll soon do) and you'll see that all eukaryotes are made of basically the same components. Most of us couldn't distinguish between a plant cell, a kidney cell and a protist from the local pond down the electron microscope: they all look remarkably similar. Just try it (**Figure 3**). If rising

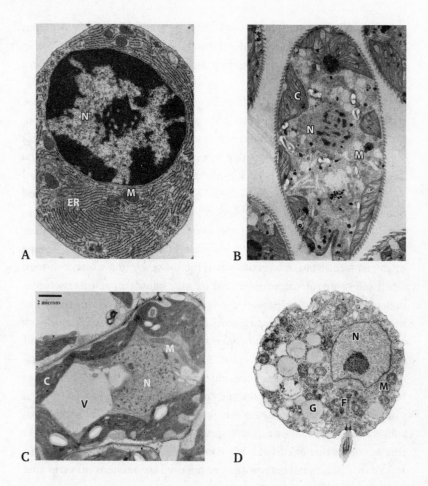

Figure 3 **The complexity of eukaryotes**
Four different eukaryotic cells showing equivalent morphological complexity. **A** shows an animal cell (a plasma cell), with a large central nucleus (N), extensive internal membranes (endoplasmic reticulum, ER) studded with ribosomes, and mitochondria (M).[Don Fawcett/Photo Researchers] **B** is the unicellular alga *Euglena*, found in many ponds, showing a central nucleus (N), chloroplasts (C), and mitochondria (M). **C** is a plant cell bounded by a cell wall, with a vacuole (V), chloroplasts (C), a nucleus (N), and mitochondria (M). **D** is a chytrid fungus zoospore, implicated in the extinction of 150 frog species; (N) is the nucleus, (M) mitochondria, (F) the flagellum, and (G) gamma bodies of unknown function.

oxygen levels removed constraints on complexity, the prediction from 'standard' natural selection is that adaptation to different ways of life in different populations should lead to a polyphyletic radiation. But that isn't what we see.

From the late 1960s, Lynn Margulis argued that this view is in any case misguided: that eukaryotic cells did not arise via standard natural selection, but through a series of endosymbioses, in which a number of bacteria cooperated together so closely that some cells physically got inside others. Such ideas trace their roots back to the early twentieth century to Richard Altmann, Konstantin Mereschkowski, George Portier, Ivan Wallin and others, who argued that all complex cells arose through symbioses between simpler cells. Their ideas were not forgotten but were laughed out of house as 'too fantastic for present mention in polite biological society'. By the time of the molecular biology revolution in the 1960s, Margulis was on firmer, albeit still controversial, ground, and we now know that at least two components of eukaryotic cells were derived from endosymbiotic bacteria – the mitochondria (the energy transducers in complex cells), which derive from α-proteobacteria; and the chloroplasts (the photosynthetic machinery of plants), deriving from cyanobacteria. Almost all the other specialised 'organelles' of eukaryotic cells have at one time or another also been claimed to be endosymbionts, including the nucleus itself, the cilia and flagella (sinuous processes whose rhythmic beat drives the movement of cells) and peroxisomes (factories for toxic metabolism). Thus the serial endosymbiosis theory claims that eukaryotes are composed of an ensemble of bacteria, forged in a communal enterprise over hundreds of millions of years after the Great Oxidation Event.

It's a poetic notion, but the serial endosymbiosis theory makes an implicit prediction equivalent to that of standard selection. If it were true, we would expect to see polyphyletic origins – a mixed bag of internal structures, as varied as the external appearance of cells. In any series of endosymbioses, where the symbiosis depends on some kind of metabolic trading in a particular environment, we would expect to find disparate types of cell interacting in different environments. If these cells later became fashioned into the organelles of complex eukaryotic cells, the hypothesis predicts that some eukaryotes should possess one set of components, and others a different set. We should expect to find all kinds of intermediates and unrelated variants lurking in obscure hiding places like stagnant muds. Right up to her prema- ture death from a stroke in 2011, Margulis did indeed hold firm to her belief that eukaryotes are a rich and varied tapestry of endosymbioses. For her, endosymbiosis was a way of life, an underexplored 'feminine' avenue of evolution, in which cooperation – 'networking', as she called it – trumped the unpleasantly masculine competition between the hunters and hunted. But in her veneration of 'real' living cells, Margulis turned her back on the more arid computational discipline of phylogenetics, the study of gene sequences and whole genomes, which has the power to tell us exactly how different eukaryotes relate to each other. And that tells a very different – and ultimately far more compelling – story.

The story hinges on a large group of species (a thousand or more in number) of simple single-celled eukaryotes that lack mitochondria. This group was once taken to be a primitive evolutionary 'missing link' between bacteria and more complex eukaryotes – exactly the kind of intermediate predicted by the serial endosymbiosis theory. The group includes the nasty intestinal parasite *Giardia*, which in Ed Yong's words resembles a malevo- lent teardrop (**Figure 4**). It lives up to its looks, causing unpleasant diarrhoea. It has not just one nucleus but two, and so is unquestionably eukaryotic, but it lacks other archetypal traits, notably mitochondria. In the mid 1980s, the iconoclastic biologist Tom Cavalier-Smith argued that *Giardia* and other relatively simple eukaryotes were probably survivors from the earliest period of eukaryotic evolution, before the acquisition of mitochondria. While Cavalier-Smith accepted that mitochondria do indeed derive from bacterial endosymbionts, he had little time for Margulis's serial endosym- biosis theory; instead, he pictured (and still does) the earliest eukaryotes as primitive phagocytes, similar to modern amoeba, which earned their living by engulfing other cells. The cells that acquired mitochondria, he argued, already had a nucleus, and a dynamic internal skeleton that helped them to change shape and move around, and protein machinery for shifting cargo about their insides, and specialised compartments for digesting food inter- nally, and so on. Acquiring mitochondria helped, certainly – they

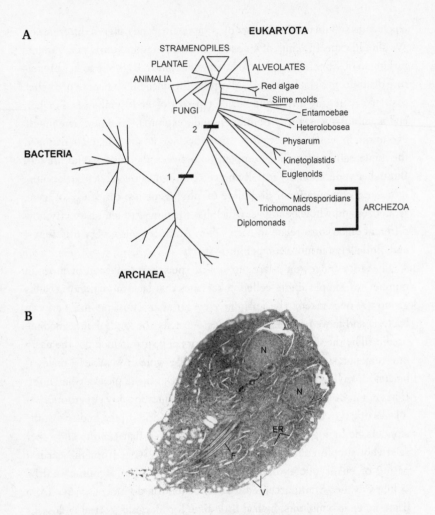

Figure 4 **The Archezoa – the fabled (but false) missing link**

A An old and misleading tree of life based on ribosomal RNA, showing the three domains of bacteria, archaea and eukaryotes. The bars mark (1) the supposed early evolution of the nucleus; and (2) the presumed later acquisition of mitochondria. The three groups that branch between the bars constitute the Archezoa, supposedly primitive eukaryotes that not yet acquired mitochondria, such as *Giardia* (**B**). We now know that the Archezoa are not primitive eukaryotes at all, but derive from more complex ancestors that already had mitochondria – they actually branch within the main part of the eukaryotic tree (N = nucleus; ER = endoplasmic reticulum; V = vacuoles; F = flagella).

turbocharged those primitive cells. But souping-up a car does not alter the structure of the car: you still begin with an automobile, which already has an engine, gearbox, brakes, everything that makes it a car. Turbocharging changes nothing but the power output. Likewise in the case of Cavalier-Smith's primitive phagocytes – everything was already in place except mitochondria, which merely gave cells more power. If there is a textbook view of eukaryotic origins – even today – this is it.

Cavalier-Smith dubbed these early eukaryotes the 'archezoa' (meaning ancient animals), to reflect their supposed antiquity (**Figure 4**). Several are parasites that cause diseases, so their biochemistry and genomes have attracted the interest of medical research and the funding that goes with it. That in turn means we now know a great deal about them. Over the past two decades, we have learned from their genome sequences and detailed biochemistry that *none* of the archezoa is a real missing link, which is to say that they are not true evolutionary intermediates. On the contrary, all of them derive from more complex eukaryotes, which once had a full quota of everything, including, and in particular, mitochondria. They had lost their erstwhile complexity while specialising to live in simpler niches. All of them retain structures that are now known to derive from mitochondria by reductive evolution – either hydrogenosomes or mitosomes. These don't look much like mitochondria, even though they have an equivalent double-membrane structure, hence the erroneous assumption that archezoa never possessed mitochondria. But the combination of molecular and phylogenetic data shows that hydrogenosomes and mitosomes are indeed derived from mitochondria, not some other bacterial endosymbiont (as predicted by Margulis). Thus all eukaryotes have mitochondria in one form or another. We can infer that the last eukaryotic common ancestor already had mitochondria, as had been predicted by Bill Martin in 1998 (see the Introduction). The fact that all eukaryotes have mitochondria may seem to be a trivial point, but when combined with the proliferation of genome sequences from across the wider microbial world, this knowledge has turned our understanding of eukaryotic evolution on its head.

We now know that eukaryotes all share a common ancestor, which by definition arose just once in the 4 billion years of life on earth. Let me reiterate this point, as it is crucial. All plants, animals, algae, fungi and protists share a common ancestor – the eukaryotes are *monophyletic*. This means that plants did not evolve from one type of bacteria, and animals or fungi from other types. On the contrary, a population of morphologically complex eukaryotic cells arose on a single occasion – and all plants, animals, algae and fungi evolved from this founder population. Any common ancestor is by definition a singular entity – not a single cell, but a single population of essentially identical cells. That does not in itself mean that the origin of complex cells was a rare event. In principle, complex cells could have arisen on numerous occasions, but only one group persisted – all the rest died out for some reason. I shall argue that this was not the case, but first we must consider the properties of eukaryotes in a little more detail.

The common ancestor of all eukaryotes quickly gave rise to five 'supergroups' with diverse cellular morphologies, most of which are obscure even to classically trained biologists. These supergroups have names like unikonts (comprising animals and fungi), excavates, chromalveolates and plantae (including land plants and algae). Their names don't matter, but two points are important. First there is far more genetic variation within each of these supergroups than there is between the ancestors of each group (**Figure 5**). That implies an explosive early radiation – specifically a *monophyletic* radiation that hints at a release from structural constraints. Second, the common ancestor was already a strikingly complex cell. By comparing traits common to each of the supergroups, we can reconstruct the likely properties of the common ancestor. Any trait present in essentially all the species of all supergroups was presumably inherited from that common ancestor, whereas any traits that are only present in one or two groups were presumably acquired later, and only in that group. Chloroplasts are a good example of the latter: they are found only in plantae and chromalveolates, as a result of well-known endosymbioses. They were not part of the eukaryotic common ancestor.

So what does phylogenetics tell us *was* part of the common ancestor? Shockingly, nearly everything else. Let me run through a few items. We know that the common ancestor had a nucleus, where it stored its DNA. The nucleus has a great deal of complex structure that is again conserved

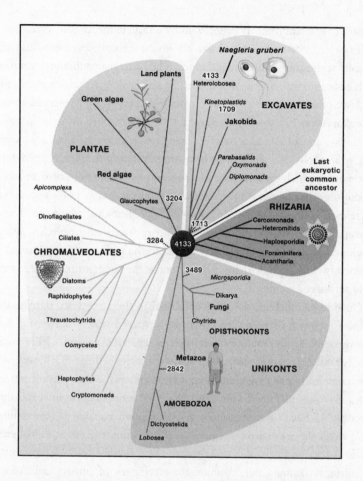

Figure 5 **The 'supergroups' of eukaryotes**
A tree of eukaryotes, based on thousands of shared genes, showing the five 'supergroups' as depicted by Eugene Koonin in 2010. The numbers refer to the number of genes shared by each of these supergroups with LECA (the last eukaryotic common ancestor). Each group has independently lost or gained many other genes. Most variation here is between single-celled protists; animals fall within the Metazoans (near the bottom). Notice that there is far more variation within each supergroup than between the ancestors of these groups, suggesting an explosive early radiation. I like the symbolic black hole at the centre: LECA had already evolved all the common eukaryotic traits, but phylogenetics gives little insight into how any of these arose from bacteria or archaea – an evolutionary black hole.

right across eukaryotes. It is enclosed by a double membrane, or rather a series of flattened sacs that look like a double membrane but are in fact continuous with other cellular membranes. The nuclear membrane is studded with elaborate protein pores and lined by an elastic matrix; and within the nucleus, other structures such as the nucleolus are again conserved across all eukaryotes. It's worth stressing that dozens of core proteins in these complexes are conserved across the supergroups, as are the histone proteins that wrap DNA. All eukaryotes have straight chromosomes, capped with 'telomeres', which prevent the ends from fraying like the tips of shoe laces. Eukaryotes have 'genes in pieces', in which short sections of DNA encoding proteins are interspersed by long non-coding regions, called introns. These introns are spliced out before being incorporated into proteins, using machinery common to all eukaryotes. Even the position of the introns is frequently conserved, with insertions found at the same position of the same gene across eukaryotes.

Outside the nucleus, the story continues in the same vein. Barring the simpler archezoa (which turn out to be scattered widely across the five supergroups, again demonstrating their independent loss of earlier complexity), all eukaryotes share essentially the same cellular machinery. All have complex internal membrane structures such as the endoplasmic reticulum and Golgi apparatus, which are specialised for packaging and exporting proteins. All have a dynamic internal cytoskeleton, capable of remodelling itself to all shapes and requirements. All have motor proteins that shuttle objects back and forth on cytoskeletal tracks across the cell. All have mitochondria, lysosomes, peroxisomes, the machinery of import and export, and common signalling systems. The list goes on. All eukaryotes divide by mitosis, in which chromosomes are separated on a microtubular spindle, using a common set of enzymes. All are sexual, with a life cycle involving meiosis (reductive division) to form gametes like the sperm and egg, followed by the fusion of these gametes. The few eukaryotes that lose their sexuality tend to fall quickly extinct (quickly in this case meaning over a few million years).

We have understood much of this for a long time from the microscopic structure of cells, but the new era of phylogenomics illuminates two aspects vividly. First, the structural similarities are not superficial resemblances, appearances that flatter to deceive, but are written out in the detailed sequences of genes, in millions and billions of letters of DNA – and that allows us to compute their ancestry as a branching tree with unprecedented precision. Second, the advent of high-throughput gene sequencing means that sampling of the natural world no longer relies on painstaking attempts to culture cells or to prepare microscopic sections, but is as fast and reliable as a shotgun sequencer. We have discovered several unexpected new groups, including eukaryotic extremophiles capable of dealing with high concentrations of toxic metals or high temperatures, and tiny but perfectly formed cells known as picoeukaryotes, as small as bacteria yet still featuring a scaled-down nucleus and midget mitochondria. All of this means we have a much clearer idea of the diversity of eukaryotes. All these new eukaryotes fit comfortably within the five established supergroups – they do not open up new phylogenetic vistas. The killer fact that emerges from this enormous diversity is how damned similar eukaryotic cells are. We do not find all kinds of intermediates and unrelated variants. The prediction of the serial endosymbiosis theory, that we should, is wrong.

That poses a different problem. The stunning success of phylogenetics and the informational approach to biology can easily blind us to its limitations. The problem here is what amounts to a phylogenetic 'event horizon' at the origin of eukaryotes. All these genomes lead back to the last common ancestor of eukaryotes, which had more or less everything. But where did all these parts come from? The eukaryotic common ancestor might as well have jumped, fully formed, like Athena from the head of Zeus. We gain little insight into traits that arose before the common ancestor – essentially all of them. How and why did the nucleus evolve? What about sex? Why do virtually all eukaryotes have two sexes? Where did the extravagant internal membranes come from? How did the cytoskeleton become so dynamic and flexible? Why does sexual cell division ('meiosis') halve chromosome numbers by first doubling them up? Why do we age, get cancer, and die? For all its ingenuity, phylogenetics can tell us little about these central questions in biology. Almost all the genes involved (encoding so-called eukaryotic 'signature proteins') are not found in prokaryotes. And conversely,

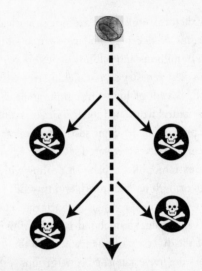

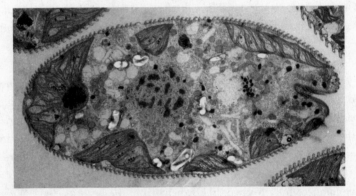

Figure 6 **The black hole at the heart of biology**
The cell at the bottom is the common alga *Euglena*, found in many ponds. You don't need to know what all these internal structures are to appreciate that this cell is large and complex. With the exception of the chloroplasts (the "leafy" membranous structures at the periphery of the cell), almost all the complex components of this cell were present in the last common ancestor of all eukaryotes. In stark contrast, at the top is a relatively complex bacterium, *Planctomycetes*, shown roughly to scale. I am not suggesting that the eukaryotes derived from *Planctomycetes* (they certainly did not), merely showing the scale of the gulf between a relatively complex bacterium and a representative single-celled eukaryote. There are no surviving evolutionary intermediates to tell the tale (indicated by the skulls and cross bones).

bacteria show practically no tendency to evolve any of these complex eukaryotic traits. There are no known evolutionary intermediates between the morphologically simple state of all prokaryotes and the disturbingly complex common ancestor of eukaryotes (**Figure 6**). All these attributes of complex life arose in a phylogenetic void, a black hole at the heart of biology.

The missing steps to complexity

Evolutionary theory makes a simple prediction. Complex traits arise via a series of small steps, each new step offering a small advantage over the last. Selection of the best-adapted traits means loss of the less well-adapted traits, so selection continuously eliminates intermediates. Over time, traits will tend to scale the peaks of an adaptive landscape, so we see the apparent perfection of eyes, but not the less perfect intermediate steps *en route* to their evolution. In *The Origin of Species* Darwin made the point that natural selection actually predicts that intermediates should be lost. In that context, it is not terribly surprising that there are no surviving intermediates between bacteria and eukaryotes. What is more surprising, though, is that the same traits do not keep on arising, time and time again – like eyes.

We do not see the historical steps in the evolution of eyes, but we do see an ecological spectrum. From a rudimentary light-sensitive spot on some early worm-like creature, eyes have arisen independently on scores of occasions. That is exactly what natural selection predicts. Each small step offers a small advantage in one particular environment, with the precise advantage depending on the precise environment. Morphologically distinct types of eye evolve in different environments, as divergent as the compound eyes of flies and mirror eyes of scallops, or as convergent as the camera eyes that are so similar in humans and octopuses. Every conceivable intermediate, from pinholes to accommodating lenses, is found in one species or another. We even see miniature eyes, replete with a 'lens' and a 'retina', in some single-celled protists. In short, evolutionary theory predicts that there should be multiple – polyphyletic – origins of traits in which each small step offers a small advantage over the last step. Theoretically that applies to all traits, and it is indeed what we generally see. So powered flight arose on at least six different occasions in bats, birds, pterosaurs and various insects; multicellularity about 30 times, as noted earlier; different forms of endo-thermy (warm blood) in several groups including mammals and birds, but also some fish, insects and plants;[6] and even conscious awareness appears to have arisen more or less independently in birds and mammals. As with eyes, we see a myriad of different forms reflecting the different environments in which they arose. Certainly there are physical constraints, but they are not strong enough to preclude multiple origins.

So what about sex, or the nucleus, or phagocytosis? The same reasoning ought to apply. If each of these traits arose by natural selection – which they undoubtedly did – and all of the adaptive steps offered some small advan-tage – which they undoubtedly did – then we should see multiple origins of eukaryotic traits in bacteria. But we don't. This is little short of an evolu-tionary 'scandal'. We see no more than the beginnings of eukaryotic traits in bacteria. Take sex, for example. Some may argue that bacteria practise a form of conjugation equivalent to sex, transferring DNA from one to another by 'lateral' gene transfer. Bacteria have all the machinery needed to recombine DNA, enabling them to forge new and varied chromosomes, which is usually taken to be the advantage of sex. But the differences are enormous. Sex involves the fusion of two gametes, each with half the normal quota of genes, followed by reciprocal recombination across the entire genome. Lateral gene transfer is neither reciprocal nor systematic in this way, but piecemeal. In effect, eukaryotes practise 'total sex', bacteria a pallid half-hearted form. Plainly there must be some advantage to eukary-otes indulging in total sex; but if so, we would expect that at least some types

––––––––––

6 The idea of endothermy in plants might seem surprising, but it is known in many different flowers, probably helping to attract pollinators by aiding the release of attractant chemicals; it may also provide a 'heat reward' for pollinating insects, promote flower development and protect against low temperatures. Some plants like the sacred lotus (*Nelumbo nucifera*) are even capable of thermoregulation, sensing changes in temperature and regulating cellular heat production to maintain tissue temperature within a narrow range.

of bacteria would do something similar, even if the detailed mechanisms were different. To the best of our knowledge, none ever did. The same goes for the nucleus and phagocytosis – and more or less all eukaryotic traits. The first steps are not the problem. We see some bacteria with folded internal membranes, others with no cell wall and a modestly dynamic cytoskeleton, yet others with straight chromosomes, or multiple copies of their genome, or giant cell size: all the beginnings of eukaryotic complexity. But bacteria always stop well short of the baroque complexity of eukaryotes, and rarely if ever combine multiple complex traits in the same cell.

The easiest explanation for the deep differences between bacteria and eukaryotes is competition. Once the first true eukaryotes had evolved, the argument goes, they were so competitive that they dominated the niche of morphological complexity. Nothing else could compete. Any bacteria that 'tried' to invade this eukaryotic niche were given short shrift by the sophisticated cells that already lived there. To use the parlance, they were outcompeted to extinction. We are all familiar with the mass extinctions of dinosaurs and other large plants and animals, so this explanation seems perfectly reasonable. The small, furry ancestors of modern mammals were held in check by the dinosaurs for millions of years, only radiating into modern groups after the dinosaurs' demise. Yet there are some good reasons to question this comfortable but deceptive idea. Microbes are not equivalent to large animals: their population sizes are enormously larger, and they pass around useful genes (such as those for antibiotic resistance) by lateral gene transfer, making them very much less vulnerable to extinction. There is no hint of any microbial extinction, even in the aftermath of the Great Oxidation Event. The 'oxygen holocaust', which supposedly wiped out most anaerobic cells, can't be traced at all: there is no evidence from either phylogenetics or geochemistry that such an extinction ever took place. On the contrary, anaerobes prospered.

More significantly, there is very strong evidence that the intermediates were not, in fact, outcompeted to extinction by more sophisticated eukaryotes. They still exist. We met them already – the 'archezoa', that large group of primitive eukaryotes that were once mistaken for a missing link. They are not true *evolutionary* intermediates, but they're real *ecological* intermediates. They occupy the same niche. An evolutionary intermediate is a missing link – a fish with legs, such as *Tiktaalik*, or a dinosaur with feathers and wings, such as *Archaeopteryx*. An ecological intermediate is not a true missing link but it proves that a certain niche, a way of life, is viable. A flying squirrel is not closely related to other flying vertebrates such as bats or birds, but it demonstrates that gliding flight between trees is possible without fully fledged wings. That means it's not pure make-believe to suggest that powered flight could have started that way. And that is the real significance of the archezoa – they are ecological intermediates, which prove that a certain way of life is viable.

I mentioned earlier that there are a thousand or more different species of archezoa. These cells are *bona fide* eukaryotes, which adapted to this 'intermediate' niche by becoming simpler, not bacteria that became slightly more complex. Let me stress the point. The niche is viable. It has been invaded on numerous occasions by morphologically simple cells, which thrive there. These simple cells were not outcompeted to extinction by more sophisticated eukaryotes that already existed and filled the same niche. Quite the reverse: they flourished precisely because they became simpler. In statistical terms, all else being equal, the probability of only simple eukaryotes (rather than complex bacteria) invading this niche on 1,000 separate occasions is about one in 10^{300} against – a number that could have been conjured up by Zaphod Beeblebrox's Infinite Improbability Drive. Even if archezoa arose independently on a far more conservative 20 separate occasions (each time radiating to produce a large number of daughter species), the probability is still one in a million against. Either this was a fluke of freakish proportions, or all else was not equal. The most plausible explanation is that there was something about the structure of eukaryotes that facilitated their invasion of this intermediate niche, and conversely, something about the structure of bacteria that precluded their evolution of greater morphological complexity.

That doesn't seem particularly radical. In fact it chimes with everything else we know. I have talked throughout this chapter about bacteria, but as we noted in the Introduction, there are actually two large groups, or *domains*, of cells that lack a nucleus, hence are designated 'prokaryotes' (literally 'before the nucleus'). These are the bacteria and the 'archaea', not to be mistaken for the archezoa, the simple eukaryotic cells that we've been discussing. While I can only apologise for the confusion of scientific terminology, which sometimes seems to be crafted by alchemists who crave not to be understood, please remember that the archaea and the bacteria are prokaryotes, lacking a nucleus, whereas the archezoa are primitive eukaryotes, which have a nucleus. In fact, the archaea are still sometimes called archaebacteria, or 'ancient bacteria', as opposed to eubacteria, or 'true bacteria', so both groups can legitimately be called bacteria. For simplicity, I'll continue to use the word bacteria loosely to refer to both groups, except when I need to specify critical differences between the two domains.[7]

The crucial point is that these two domains, the bacteria and the archaea, are extremely different in their genetics and in their biochemistry, but almost indistinguishable in their morphology. Both types are small simple cells that lack a nucleus and all the other eukaryotic traits that define complex life. The fact that both groups failed to evolve complex morphology, despite their extraordinary genetic diversity and biochemical ingenuity, makes it look as if an intrinsic physical constraint precludes the evolution of complexity in prokaryotes, a constraint that was somehow released in the evolution of eukaryotes. In Chapter 5, I'll argue that this constraint was released by a rare event – the singular endosymbiosis between two prokaryotes that we discussed in the Introduction. For now, though, let's just note that some sort of structural constraint must have acted equally on both of the two great domains of prokaryotes, the bacteria and archaea, forcing both groups to remain simple in their morphology throughout an incomprehensible 4 billion years. Only eukaryotes explored the realm of complexity, and they did so via an explosive monophyletic radiation that implies a release from whatever these structural constraints might have been. That appears to have happened just once – all eukaryotes are related.

---

7 All of these words are heavily loaded with intellectual and emotional baggage, accumulated over decades. The terms archaebacteria and archaea are technically incorrect anyway as the domain is no older than the bacteria. I prefer to use the terms archaea and bacteria, in part because they emphasise the surprisingly fundamental differences between the two domains, and in part because they are just simpler.

The wrong question

This, then, is our short history of life through new eyes. Here is a swift summary. The early earth was not drastically different from our own world: it was a water world, with a moderate climate, dominated by volcanic gases such as carbon dioxide and nitrogen. While our early planet lacked oxygen, it was not rich in gases conducive to organic chemistry – hydrogen, methane and ammonia. That rules out tired old ideas of primordial soup; yet life started as early as could be, perhaps 4 billion years ago. At face value, something else was driving the emergence of life; we will come to that. Bacteria soon took over, colonising every inch, every metabolic niche, remodelling the globe over 2 billion years, depositing rocks and minerals on a colossal scale, transforming oceans, atmosphere and continents. They crashed the climate in global snowball earths; they oxidised the world, filling the oceans and air with reactive oxygen. Yet in all this immense duration, neither the bacteria nor the archaea became anything else: they remained stubbornly simple in their structure and way of life. For an eternal 4 billion years, through extremes of environmental and ecological change, bacteria changed their genes and biochemistry, but never changed their form. They never gave rise to more complex life forms, of the kind we might hope to detect on another planet, intelligent aliens – except just once.

On one single occasion, here on earth, bacteria gave rise to eukaryotes. There is nothing in the fossil record, or in phylogenetics, to suggest that complex life actually arose repeatedly, but that only one group, the familiar modern eukaryotes, survived. On the contrary, the monophyletic radiation of eukaryotes suggests their unique origin was dictated by intrinsic physical constraints which had little if anything to do with environmental upheavals such as the Great Oxidation Event. We'll see what these constraints might have been in Part III. For now, let's just note that any proper account must explain why the evolution of complex life happened only once: our explanation must be persuasive enough to be believable, but not so persuasive that we are left wondering why it did not happen on many occasions. Any attempt to explain a singular event will always have the appearance of a fluke about it. How can we prove it one way or another? There might not be much to go on in the event itself, but there may be clues concealed in the aftermath, a smoking gun that gives some indication of what happened. Once they cast off their bacterial shackles, the eukaryotes became enormously complex and diverse in their morphology. Yet they did not accrue this complexity in an obviously predictable way: they came up with a whole series of traits, from sex and ageing to speciation, none of which have ever been seen in bacteria or archaea. The earliest eukaryotes accumulated all these singular traits in a common ancestor without peer. There are no known evolutionary intermediates between the morphological simplicity of bacteria and that enormously complex eukaryotic common ancestor to tell the tale. All of this adds up to a thrilling prospect – the biggest questions in biology remain to be solved! Is there some pattern to these traits that might give an indication of how they evolved? I think so.

This puzzle relates back to the question that we asked at the beginning of this chapter. How much of life's history and properties can be predicted from first principles? I suggested that life is constrained in ways that can't easily be interpreted in terms of genomes, history or environment. If we consider life in terms of information alone, my contention was that we could predict none of this inscrutable history. Why did life start so early? Why did it stagnate in morphological structure for billions of years? Why were bacteria and archaea unaffected by environmental and ecological upheavals on a global scale? Why is all complex life monophyletic, arising just once in 4 billion years? Why do prokaryotes not continuously, or even occasionally, give rise to cells and organisms with greater complexity? Why do individual eukaryotic traits such as sex, the nucleus and phagocytosis not arise in bacteria or archaea? Why did eukaryotes accumulate all these traits?

If life is all about information, these are deep mysteries. I don't believe this story could be foretold, predicted as science, on the basis of information alone. The quirky properties of life would have to be ascribed to the contingencies of history, the slings and arrows of outrageous fortune. We would have no possibility of predicting the properties of life on other planets. Yet DNA, the beguiling code-script which seems to promise every answer, has made us forget Schrödinger's other central tenet – that life resists entropy, the tendency to decay. In a footnote to *What is Life?* Schrödinger noted that if he had been writing for an audience of physicists, he would have framed his argument not in terms of entropy, but of free energy. That word 'free' has a specific meaning, which we will consider in the next chapter; suffice for now to say that energy is precisely what was missing from this chapter, and indeed from Schrödinger's book. His iconic title asked the wrong question altogether. Add in energy, and the question is much more telling: *What is Living?* But Schrödinger must be forgiven. He could not have known. When he was writing, nobody knew much about the biological currency of energy. Now we know how it all works in exquisite detail, right down to the level of atoms. The detailed mechanisms of energy harvesting turn out to be conserved as universally across life as the genetic code itself, and these mechanisms exert fundamental structural constraints on cells. But we have no idea how they evolved, nor how biological energy constrained the story of life. That is the question of this book.

2

WHAT IS LIVING?

I t is a cold killer, with a calculated cunning honed over millions of generations. It can interfere with the sophisticated immune surveillance machinery of an organism, melting unobtrusively into the background like a double
agent. It can recognise proteins on the cell surface, and lock on to them as
if it were an insider, gaining entrance to the inner sanctum. It can home in
unerringly on the nucleus, and incorporate itself into a host cell's DNA.
Sometimes it remains there in hiding for years, invisible to all around. On
other occasions it takes over without delay, sabotaging the host cell's biochemical machinery, making thousands upon thousands of copies of itself.
It dresses up these copies in a camouflaged tunic of lipids and proteins, ships
them to the surface, and bursts out to begin another round of guile and
destruction. It can kill a human being cell by cell, person by person, in devastating epidemic, or dissolve entire oceanic blooms extending over hundreds of miles, overnight. Yet most biologists would not even classify it as
alive. The virus itself doesn't give a damn.

Why would a virus not be alive? Because it does not have any active
metabolism of its own; it relies entirely on the power of its host. That raises
the question – is metabolic activity a necessary attribute of life? The pat
answer is yes, of course; but why, exactly? Viruses use their immediate
environment to make copies of themselves. But then so do we: we eat other
animals or plants, and we breathe in oxygen. Cut us off from our

environment, say with a plastic bag over the head, and we die in a few minutes. One could say that we parasitise our environment – like viruses. So do plants. Plants need us nearly as much as we need them. To photosynthesise their own organic matter, to grow, plants need sunlight, water and carbon dioxide (CO_2). Arid deserts or dark caves preclude growth, but so too would a shortage of CO_2. Plants are not short of the gas precisely because animals (and fungi and various bacteria) continuously break down organic matter, digesting it, burning it, finally releasing it back into the atmosphere as CO_2. Our extra efforts to burn all the fossil fuels may have horrible consequences for the planet, but plants have good cause to be grateful. For them, more CO_2 means more growth. So, like us, plants are parasites of their environment.

From this point of view, the difference between plants and animals and viruses is little more than the largesse of that environment. Within our cells, viruses are cosseted in the richest imaginable womb, a world that provides their every last want. They can afford to be so pared down – what Peter Medawar once called 'a piece of bad news wrapped in a protein coat' – only because their immediate environment is so rich. At the other extreme, plants place very low demands on their immediate environment. They will grow almost anywhere with light, water and air. To eke out an existence with so few external requirements forces them to be internally sophisticated. In terms of their biochemistry, plants can produce everything they need to grow, literally synthesising it out of thin air.[1] We ourselves are somewhere in the middle. Beyond a general requirement for eating, we need specific vitamins in our diet, without which we succumb to nasty diseases like scurvy. Vitamins are compounds that we can't make

---

1 They also need minerals such as nitrate and phosphate, of course. Many cyanobacteria (the bacterial precursors of plants' photosynthetic organelles, the chloroplasts) can fix nitrogen, which is to say, they can convert the relatively inert nitrogen gas (N_2) in the air into the more active and usable form ammonia. Plants have lost this ability, and rely on the largesse of their environment, sometimes in the form of symbiotic bacteria in the root nodules of leguminous plants, which provide their active nitrogen. Without that extraneous biochemical machinery, plants, like viruses, could not grow or reproduce. Parasites!

for ourselves from simple precursors, because we have lost our ancestors' biochemical machinery for synthesising them from scratch. Without the external props provided by vitamins, we are as doomed as a virus without a host.

So we all need props from the environment, the only question is how many? Viruses are actually extremely sophisticated in relation to some parasites of DNA, such as retrotransposons (jumping genes) and the like. These never leave the safety of their host, yet copy themselves across whole genomes. Plasmids – typically small independent rings of DNA carrying a handful of genes – can pass directly from one bacterium to another (via a slender connecting tube) without any need to fortify themselves to the outer world. Are retrotransposons, plasmids and viruses alive? All share a kind of 'purposeful' cunning, an ability to take advantage of their immediate biological environment, to make copies of themselves. Plainly there is a continuum between non-living and living, and it is pointless to try to draw a line across it. Most definitions of life focus on the living organism itself, and tend to ignore life's parasitising of its environment. Take the NASA 'working definition' of life, for example: life is 'a self-sustaining chemical system capable of Darwinian evolution'. Does that include viruses? Probably not, but it depends on what we read into that slippery phrase 'self-sustaining'. Either way, life's dependence on its environment is not exactly emphasised. The environment, by its very nature, seems extraneous to life; we shall see that it is not at all. The two always go hand in hand.

What happens when life is cut off from its preferred environment? We die, of course: we are either living or dead. But that's not always true. When cut off from the resources of a host cell, viruses do not instantly decay and 'die': they are fairly impervious to the depredations of the world. In every millilitre of seawater, there are ten times as many viruses, waiting for their moment, as there are bacteria. A virus's resistance to decay is reminiscent of a bacterial spore, which is held in a state of suspended animation and can remain that way for many years. Spores survive thousands of years in permafrost, or even in outer space, without metabolising at all. They're not alone: seeds and even animals like tardigrades can withstand extreme conditions such as complete dehydration, radiation a thousand times the dose that would kill a human, intense pressures at the bottom of the ocean, or the vacuum of space – all without food or water.

Why do viruses, spores and tardigrades not fall to pieces, conforming to the universal decay dictated by the second law of thermodynamics? They might do in the end – if frazzled by the direct hit of a cosmic ray or a bus – but otherwise they are almost completely stable in their non-living state. That tells us something important about the difference between life and living. Spores are not technically living, even though most biologists would classify them as alive, because they retain the potential to revive. They can go back to living, so they're not dead. I don't see why we should view viruses in any different light: they too revert to copying themselves as soon as they are in the right environment. Likewise tardigrades. Life is about its structure (dictated in part by genes and evolution), but living – growing, proliferating – is as much about the environment, about how structure and environment interrelate. We know a tremendous amount about how genes encode the physical components of cells, but far less about how physical constraints dictate the structure and evolution of cells.

Energy, entropy and structure

The second law of thermodynamics states that entropy – disorder – must increase, so it seems odd at first glance that a spore or a virus should be so stable. Entropy, unlike life, has a specific definition and can be measured (it has units of joules per kelvin per mole, since you ask). Take a spore and smash it to smithereens: grind it up into all of its molecular components, and measure the entropy change. Surely entropy must have increased! What was once a beautifully ordered system, capable of resuming growth as soon as it found suitable conditions, is now a random non-functional assortment of bits – high entropy by definition. But no! According to the careful measurements of the bioenergeticist Ted Battley, entropy barely changed. That's because there is more to entropy than just the spore; we must also consider its surroundings, and they have some level of disorder too.

A spore is composed of interacting parts that fit snugly together. Oily (lipid) membranes partition themselves from water naturally because of physical forces acting between molecules. A mixture of oily lipids shaken up in water will spontaneously sort itself into a thin bilayer, a biological membrane enclosing a watery vesicle, because that is the most stable state (**Figure 7**). For related reasons, an oil slick will spread into a thin layer across the surface of the ocean, causing devastation to life over hundreds of square miles. Oil and water are said to be immiscible – the physical forces of attraction and repulsion mean they prefer to interact with themselves rather than each other. Proteins behave in much the same way: those with lots of electrical charges dissolve in water; those without charges interact much better with oils – they are hydrophobic, literally 'water-hating'. When oily molecules nestle down together, and electrically charged proteins dissolve in water, energy is released: that is a physically stable, low-energy, 'comfortable' state of matter. Energy is released as heat. Heat is the motion of molecules, their jostling and molecular disorder. Entropy. So the release of heat when oil and water separate actually increases entropy. In terms of *overall* entropy, then, and taking all these physical interactions into consideration, an ordered oily membrane around a cell is a *higher* entropy state than a random mixture of immiscible molecules, even though it *looks* more ordered.[2]

Grind up a spore and the overall entropy hardly changes, because although the crushed spore itself is more disordered, the component parts now have a higher energy than they did before – oils are mixed with water, immiscible proteins are rammed hard together. This physically 'uncomfortable' state costs energy. If a physically comfortable state releases energy into the surroundings as heat, a physically uncomfortable state does the opposite. Energy has to be absorbed from the surroundings, lowering their entropy, cooling them down. Writers of horror stories grasp the central point in their chilling narratives – almost literally. Spectres, poltergeists and Dementors chill, or even freeze, their immediate surroundings, sucking out energy to pay for their unnatural existence.

---

2 Something similar happens when a star forms: here, the physical force of gravitation acting between matter offsets the local loss of disorder, but the huge release of heat produced by nuclear fusion increases disorder elsewhere in the solar system and the universe.

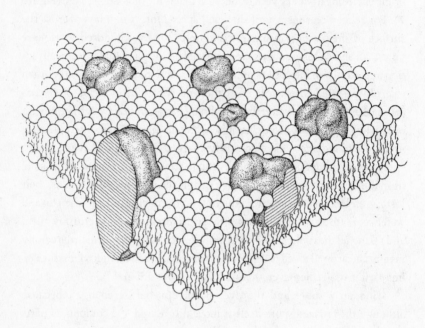

Figure 7 **Structure of a lipid membrane**
The original fluid-mosaic model of the lipid bilayer, as depicted by Singer and
Nicholson in 1972. Proteins float, submerged in a sea of lipids, some partially
embedded, and others extending across the entire membrane. The lipids
themselves are composed of hydrophilic (water-loving) head-groups, typically
glycerol phosphate, and hydrophobic (water-hating) tails, generally fatty acids in
bacteria and eukaryotes. The membrane is organised as a bilayer, with
hydrophilic heads interacting with the watery contents of the cytoplasm and the
surroundings, and the hydrophobic tails pointing inwards and interacting with
each other. This is a low-energy, physically 'comfortable' state: despite its
ordered appearance, the formation of lipid bilayers actually increases overall
entropy by releasing energy as heat into the surroundings.

When all this is taken into consideration in the case of the spore, the overall entropy barely changes. At the molecular level, the structure of polymers minimises energy locally, with excess energy being released as heat to the surroundings, increasing their entropy. Proteins naturally fold into shapes with the lowest possible energy. Their hydrophobic parts are buried far from water at the surface. Electrical charges attract or repel each other: positive charges are fixed in their place by counterbalancing negative charges, stabilising the three-dimensional structure of the protein. So proteins fold spontaneously into particular shapes, albeit not always in a helpful manner. Prions are perfectly normal proteins that spontaneously refold into semicrystalline structures that act as a template for more refolded prions. The overall entropy barely changes. There may be several stable states for a protein, only one of them useful for a cell; but in terms of entropy there is little difference between them. Perhaps most surprisingly, there is little difference in overall entropy between a disordered soup of individual amino acids (the building blocks of proteins) and a beautifully folded protein. To unfold the protein returns it to a state more similar to a soup of amino acids, increasing its entropy. But doing so also exposes the hydrophobic amino acids to water, and this physically uncomfortable state sucks in energy from outside, decreasing the entropy of the surroundings, cooling them down – what we might call the 'poltergeist effect'. The idea that life is a low-entropy state – that it is more organised than a soup – is not strictly true. The order and organisation of life is more than matched by the increased disorder of the surroundings.

So what was Erwin Schrödinger talking about when he said that life 'sucks' negative entropy from its environment, by which he meant that life somehow extracts order from its surroundings. Well, even though a broth of amino acids might have the same entropy as a perfectly folded protein, there are two senses in which the protein is less probable, and therefore costs energy.

First, the broth of amino acids will not spontaneously join together to form a chain. Proteins are chains of linked amino acids, but the amino acids are not intrinsically reactive. To get them to join together, living cells need first to activate them. Only then will they react and form into a chain. This releases roughly the same amount of energy that was used to activate them in the first place, so the overall entropy in fact remains about the same. The energy released as the protein folds itself is lost as heat, increasing the entropy of the surroundings. And so there is an *energy barrier* between two equivalently stable states. Just as the energy barrier means it is tricky to get proteins to form, so too there is a barrier to their degradation. It takes some effort (and digestive enzymes) to break proteins back down into their component parts. We must appreciate that the tendency of organic molecules to interact with each other, to form larger structures, whether proteins, DNA or membranes, is no more mysterious than the tendency of large crystals to form in cooling lava. Given enough *reactive* building blocks, these larger structures are the most stable state. The real question is: where do all the reactive building blocks come from?

That brings us to the second problem. A broth of amino acids, let alone activated ones, is not exactly probable in today's environment either. If left standing around, it will eventually react with oxygen and revert to a simpler mixture of gases – carbon dioxide, nitrogen and sulphur oxides, and water vapour. In other words, it takes energy to form these amino acids in the first place, and that energy is released when they are broken down again. That's why we can survive starvation for a while, by breaking down the protein in our muscles and using it as a fuel. This energy does not come from the protein itself, but from burning up its constituent amino acids. Thus, seeds, spores and viruses are not perfectly stable in today's oxygen-rich environment. Their components will react with oxygen – oxidise – slowly over time, and that ultimately erodes their structure and function, preventing them from springing back to life in the right conditions. Seeds die. But change the atmosphere, keep oxygen at bay, and they are stable indefinitely.[3]

---

3 A more human example is the Vasa, a formidable seventeenth-century Swedish warship that sank in the bay outside Stockholm on its maiden voyage in 1628, and was salvaged in 1961. It had been wonderfully preserved as the growing city of Stockholm poured its sewage into the marine basin. It was literally preserved in shit, with the sewer gas hydrogen sulphide preventing oxygen from attacking the ship's exquisite wooden carving. Ever since raising the ship, it has been a fight to keep it intact.

Because organisms are 'out of equilibrium' with the oxygenated global environment, they will tend to oxidise, unless the process is actively prevented. (We'll see in the next chapter that this was not always the case.)

So under normal circumstances (in 'the presence of oxygen) it costs energy to make amino acids and other biological building blocks, such as nucleotides, from simple molecules like carbon dioxide and hydrogen. And it costs energy to join them up into long chains, polymers such as proteins and DNA, even though there is little change in entropy. That's what living is all about – making new components, joining them all together, growing, reproducing. Growth also means actively transporting materials in and out of the cell. All of this requires a continuous flux of energy – what Schrödinger referred to as 'free energy'. The equation he had in mind is an iconic one, which relates entropy and heat to free energy. It is simple enough:

$$\Delta G = \Delta H - T\Delta S.$$

What does this mean? The Greek symbol Δ (delta) signifies a change. ΔG is the change in Gibbs free energy, named after the great reclusive nineteenth-century American physicist J. Willard Gibbs. It is the energy that is 'free' to drive mechanical work such as muscle contraction or anything happening in the cell. ΔH is the change in heat, which is released into the surroundings, warming them up, and so increasing their entropy. A reaction that releases heat into the surroundings must cool the system itself, because there is now less energy in it than there was before the reaction. So if heat is released from the system into the surroundings, ΔH, which refers to the system, takes a negative sign. T is the temperature. It matters just for context. Releasing a fixed amount of heat into a cool environment has a greater effect on that environment than the exact same amount of heat released into a warm environment – the relative input is greater. Finally, ΔS is the change in entropy of the system. This takes a negative sign if the entropy of the system decreases, becoming more ordered, and is positive if entropy increases, with the system becoming more chaotic.

Overall, for any reaction to take place spontaneously the free energy, ΔG, must be negative. This is equally true for the total sum of all the reactions that constitute living. That's to say, a reaction will take place on its own accord *only* if ΔG is negative. For that to be the case, either the entropy of the system must rise (the system becomes more disordered) or energy must be lost from the system as heat, or both. This means that local entropy can decrease – the system can become more ordered – so long as ΔH is even more negative, meaning that a lot of heat is released to the surroundings. The bottom line is that, to drive growth and reproduction – living! – some reaction must continuously release heat into the surroundings, making them more disordered. Just think of the stars. They pay for their ordered existence by releasing vast amounts of energy into the universe. In our own case, we pay for our continued existence by releasing heat from the unceasing reaction that is respiration. We are continuously burning food in oxygen, releasing heat into the environment. That heat loss is not waste – it is strictly necessary for life to exist. The greater the heat loss, the greater the possible complexity.[4]

Everything that happens in a living cell is spontaneous, and will take place on its own accord, given the right starting point. ΔG is always negative. Energetically, it's downhill all the way. But this means that the starting point has to be very high up. To make a protein, the starting point is the improbable assembly of enough *activated* amino acids in a small space. They will then release energy when they join and fold to form proteins, increasing the entropy of the surroundings. Even the activated amino acids will form spontaneously, given enough suitably reactive precursors. And those suitably reactive precursors will also form spontaneously, *given a highly reactive environment.* Thus, ultimately, the power for growth comes from the reactivity of the environment, which fluxes continuously through living cells (in the form of food and oxygen in our case, photons of light in the case of plants). Living cells couple this continuous energy flux to

---

4 That's an interesting point in terms of the evolution of endothermy, or hot blood. While there is no necessary connection between the greater heat loss of endotherms and greater complexity, it is nonetheless true that greater complexity must ultimately be paid for by greater heat loss. Thus endotherms could in principle (even if they don't in fact) attain greater complexity than ectotherms. Perhaps the sophisticated brains of some birds and mammals are a case in point.

growth, overcoming their tendency to break down again. They do so through ingenious structures, in part specified by genes. But whatever those structures may be (we'll come to that), they are themselves the outcome of growth and replication, natural selection and evolution, none of which is possible in the absence of a continuous energy flux from somewhere in the environment.

The curiously narrow range of biological energy

Organisms require an extraordinary amount of energy to live. The energy 'currency' used by all living cells is a molecule called ATP, which stands for adenosine triphosphate (but don't worry about that). ATP works like a coin in a slot machine. It powers one turn on a machine that promptly shuts down again afterwards. In the case of ATP, the 'machine' is typically a protein. ATP powers a change from one stable state to another, like flipping a switch from up to down. In the case of the protein, the switch is from one stable conformation to another. To flip it back again requires another ATP, just as you have to insert another coin in the slot machine to have a second go. Picture the cell as a giant amusement arcade, filled with protein machinery, all powered by ATP coins in this way. A single cell consumes around 10 *million* molecules of ATP every second! The number is breathtaking. There are about 40 trillion cells in the human body, giving a total turnover of ATP of around 60–100 kilograms per day – roughly our own body weight. In fact, we contain only about 60 grams of ATP, so we know that every molecule of ATP is recharged once or twice a minute.

Recharged? When ATP is 'split', it releases free energy that powers the conformational change, as well as releasing enough heat to keep ΔG negative. ATP is usually split into two unequal pieces, ADP (adenosine diphosphate) and inorganic phosphate (PO_4^{3-}). This is the same stuff we use in fertilisers and is usually depicted as P_i. It then costs energy to reform ATP again from ADP and P_i. The energy of respiration – the energy released from the reaction of food with oxygen – is used to make ATP from ADP and P_i. That's it. The endless cycle is as simple as this:

$$ADP + P_i + energy \rightleftharpoons ATP$$

We are nothing special. Bacteria such as *E. coli* can divide every 20 minutes. To fuel its growth *E. coli* consumes around 50 *billion* ATPs per cell division, some 50–100 times each cell's mass. That's about four times our own rate of ATP synthesis. Convert these numbers into power measured in watts and they are just as incredible. We use about 2 milliwatts of energy per gram – or some 130 watts for an average person weighing 65 kg, a bit more than a standard 100 watt light bulb. That may not sound like a lot, but per gram it is a factor of 10,000 more than the sun (only a tiny fraction of which, at any one moment, is undergoing nuclear fusion). Life is not much like a candle; more of a rocket launcher.

From a theoretical point of view, then, life is no mystery. It doesn't contravene any laws of nature. The amount of energy that living cells get through, second by second, is astronomical, but then the amount of energy pouring in upon the earth as sunlight is many orders of magnitude more (because the sun is enormously larger, even though it has less power per gram). So long as some portion of this energy is available to drive biochemistry, one might think that life could operate in almost any which way. As we saw with genetic information in the last chapter, there doesn't seem to be any fundamental constraint on how energy is used, just that there be plenty of it. That makes it all the more surprising, then, that life on earth turns out to be extremely constrained in its energetics.

There are two aspects to the energy of life that are unexpected. First, all cells derive their energy from just one particular type of chemical reaction known as a *redox* reaction, in which electrons are transferred from one molecule to another. Redox stands for 'reduction and oxidation'. It is simply the transfer of one or more electrons from a donor to a receptor. As the donor passes on electrons, it is said to be oxidised. This is what happens when substances such as iron react with oxygen – they pass electrons on to oxygen, themselves becoming oxidised to rust. The substance that receives the electrons, in this case oxygen, is said to be reduced. In respiration or a fire, oxygen (O_2) is reduced to water (H_2O) because each oxygen atom picks up two electrons (to give O^{2-}) plus two protons, which balance the charges.

The reaction proceeds because it releases energy as heat, increasing entropy. All chemistry ultimately increases the heat of the surroundings and lowers the energy of the system itself; the reaction of iron or food with oxygen does that particularly well, releasing a large amount of energy (as in a fire). Respiration *conserves* some of the energy released from that reaction in the form of ATP, at least for the short period until ATP is split again. That releases the remaining energy contained in the $ADP-P_i$ bond of ATP as heat. In the end, respiration and burning are equivalent; the slight delay in the middle is what we know as life.

Because electrons and protons are often (but not always) coupled together in this way, reductions are sometimes defined as the transfer of a hydrogen atom. But reductions are much easier to grasp if you think primarily in terms of electrons. A sequence of oxidation and reduction (redox) reactions amounts to the transfer of an electron down a linked chain of carriers, which is not unlike the flow of electrical current down a wire. This is what happens in respiration. Electrons stripped from food are not passed directly to oxygen (which would release all the energy in one go) but to a 'stepping stone' – typically one of several charged iron atoms (Fe^{3+}) embedded in a respiratory protein, often as part of a small inorganic crystal known as an 'iron–sulphur cluster' (see **Figure 8**). From there the electron hops to a very similar cluster, but with a slightly higher 'need' for the electron. As the electron is drawn from one cluster to the next, each one is first reduced (accepting an electron so an Fe^{3+} becomes Fe^{2+}) and then oxidised (losing the electron and reverting to Fe^{3+}) in turn. Ultimately, after around 15 or more such hops, the electron reaches oxygen. Forms of growth that at first glance seem to have little in common, such as photosynthesis in plants, and respiration in animals, turn out to be basically the same in that they both involve the transfer of electrons down such 'respiratory chains'. Why should this be? Life could have been driven by thermal or mechanical energy, or radioactivity, or electrical discharges, or UV radiation, the imagination is the limit; but no, all life is driven by redox chemistry, via remarkably similar respiratory chains.

The second unexpected aspect to the energy of life is the detailed mechanism by which energy is conserved in the bonds of ATP. Life doesn't use

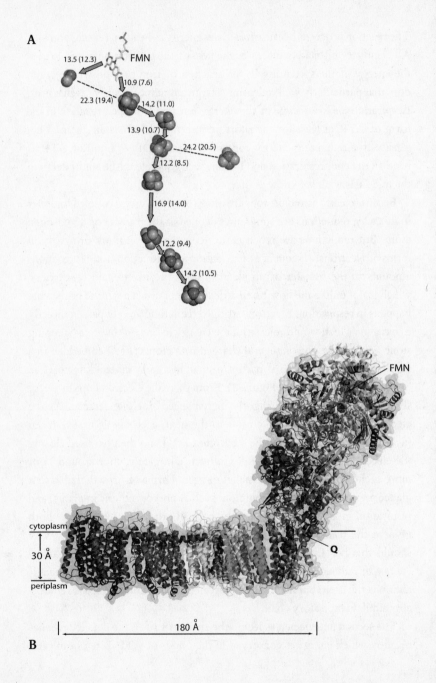

A

13.5 (12.3)

FMN

22.3 (19.4)

10.9 (7.6)

14.2 (11.0)

13.9 (10.7)

24.2 (20.5)

12.2 (8.5)

16.9 (14.0)

12.2 (9.4)

14.2 (10.5)

FMN

cytoplasm

30 Å

periplasm

Q

180 Å

B

C

Figure 8 **Complex I of the respiratory chain**
A Iron-sulphur clusters are spaced at regular distances of 14 ångströms or less; electrons hop from one cluster to the next by 'quantum tunnelling', with most following the main path of the arrows. The numbers give the distance in ångströms from centre to centre of each cluster; the numbers in brackets give the distance from edge to edge. **B** The whole of complex I in bacteria in Leo Sazanov's beautiful X-ray crystallography structure. The vertical matrix arm transfers electrons from FMN, where they enter the respiratory chain, to coenzyme Q (also called ubiquinone), which passes them on to the next giant protein complex. You can just make out the pathway of iron-sulphur clusters shown in **A** buried within the protein. Reproduced courtesy of Nature Publishing Group. **C** Mammalian complex I, showing the same core subunits found in bacteria, but partially concealed beneath an additional 30 smaller subunits, depicted in dark shades in Judy Hirst's revealing electron cryo-microscopy structure.

plain chemistry, but drives the formation of ATP by the intermediary of proton gradients across thin membranes. We'll come to what that means, and how it is done, in a moment. For now, let's just recall that this peculiar mechanism was utterly unanticipated – 'the most counterintuitive idea in biology since Darwin', according to the molecular biologist Leslie Orgel. Today, we know the molecular mechanisms of how proton gradients are generated and tapped in astonishing detail. We also know that the use of proton gradients is universal across life on earth – proton power is as much part and parcel of life as DNA itself, the universal genetic code. Yet we know next to nothing about how this counterintuitive mechanism of biological energy generation evolved. For whatever reason, it seems that life on earth uses a startlingly limited and strange subset of possible energetic mechanisms. Does this reflect the quirks of history, or are these methods so much better than anything else that they eventually came to dominate? Or more intriguingly – could this be the only way?

Here's what is happening in you right now. Take a dizzying ride down into one of your cells, let's say a heart muscle cell. Its rhythmic contractions are powered by ATP, which is flooding out from the many large mitochondria, the powerhouses of the cell. Shrink yourself down to the size of an ATP molecule, and zoom in through a large protein pore in the external membrane of a mitochondrion. We find ourselves in a confined space, like the engine room of a boat, packed with overheating protein machinery, stretching as far as the eye can see. The ground is bubbling with what look like little balls, which shoot out from the machines, appearing and disappearing in milliseconds. Protons! This whole space is dancing with the fleeting apparitions of protons, the positively charged nuclei of hydrogen atoms. No wonder you can barely see them! Sneak through one of those monstrous protein machines into the inner bastion, the matrix, and an extraordinary sight greets you. You are in a cavernous space, a dizzying vortex where fluid walls sweep past you in all directions, all jammed with gigantic clanking and spinning machines. Watch your head! These vast protein complexes are sunk deeply into the walls, and move around sluggishly as if submerged in the sea. But their parts move at amazing speed. Some sweep back and forth, too fast for the eye to see, like the pistons of a steam engine. Others spin on

their axis, threatening to detach and fly off at any moment, driven by pirou-
etting crankshafts. Tens of thousands of these crazy perpetual motion
machines stretch off in all directions, whirring away, all sound and fury,
signifying ... what?

You are at the thermodynamic epicentre of the cell, the site of cellular
respiration, deep within the mitochondria. Hydrogen is being stripped from
the molecular remains of your food, and passed into the first and largest of
these giant respiratory complexes, complex I. This great complex is composed of as many as 45 separate proteins, each one a chain of several
hundred amino acids. If you, an ATP, were as big as a person, complex I is
a skyscraper. But no ordinary skyscraper – a dynamic machine operating
like a steam engine, a terrifying contraption with a life of its own. Electrons
are separated from protons and fed into this vast complex, sucked in at one
end and spat out of the other, all the way over there, deep in the membrane
itself. From there the electrons pass through two more giant protein complexes, which together comprise the respiratory chain. Each individual
complex contains multiple 'redox centres' – about nine of them in complex
I – that transiently hold an electron (**Figure 8**). Electrons hop from centre
to centre. In fact, the regular spacing of these centres suggests that they
'tunnel' by some form of quantum magic, appearing and disappearing fleet-
ingly, according to the rules of quantum probability. All that the electrons
can see is the next redox centre, so long as it is not far away. Distance here
is measured in ångströms (Å), roughly the size of an atom.[5] So long as each
redox centre is spaced within about 14 Å of the next, and each one has a

---

5 1 ångström (Å) is 10^{-10} m, or one ten-billionth of a metre. It's technically an outmoded
term now, generally replaced by the nanometre (nm), which is 10^{-9} m, but it is still
very useful for considering distances across proteins. 14 Å is 1.4 nm. Most of the redox
centres in the respiratory chain are between 7 and 14 Å apart, with a few stretching out to
18 Å. To say that they are between 0.7 and 1.4 nm apart is the same thing, but somehow
compresses our sense of that range. The inner mitochondrial membrane is 60 Å across
– a deep ocean of lipids compared with a flimsy 6 nm! Units do condition our sense of
distance.

slightly stronger affinity for an electron than the last, electrons will hop on down this pathway of redox centres, as if crossing a river on nice regularly spaced stepping stones. They pass straight through the three giant respiratory complexes, but don't notice them any more than you need to notice the river. They are drawn onwards by the powerful tug of oxygen, its voracious chemical appetite for electrons. This is not action at a distance – it is all about the probability of an electron being on oxygen rather than somewhere else. It amounts to a wire, insulated by proteins and lipids, channelling the current of electrons from 'food' to oxygen. Welcome to the respiratory chain!

The electrical current animates everything here. The electrons hop along their path, interested only in their route to oxygen, and oblivious to the clanking machines clinging to the landscape like pumpjack oil wells. But the giant protein complexes are full of trip switches. If an electron sits in a redox centre, the adjoining protein has a particular structure. When that electron moves on, the structure shifts a fraction, a negative charge readjusts itself, a positive charge follows suit, whole networks of weak bonds recalibrate themselves, and the great edifice swings into a new conformation in a tiny fraction of a second. Small changes in one place open cavernous channels elsewhere in the protein. Then another electron arrives, and the entire machine swings back to its former state. The process is repeated tens of times a second. A great deal is known now about the structure of these respiratory complexes, down to a resolution of just a few ångströms, nearly the level of atoms. We know how protons bind to immobilised water molecules, themselves pinioned in their place by charges on the protein. We know how these water molecules shift when the channels reconfigure themselves. We know how protons are passed from one water molecule to another through dynamic clefts, opening and closing in swift succession, a perilous route through the protein that slams closed instantly after the passage of the proton, preventing its retreat as if in an Indiana Jones adventure, the Proteins of Doom. This vast, elaborate, mobile machinery achieves just one thing: it transfers protons from one side of the membrane to the other.

For each pair of electrons that passes through the first complex of the respiratory chain, four protons cross the membrane. The electron pair then passes directly into the second complex (technically complex III; complex II is an alternative entry point), which conveys four more protons across the barrier. Finally, in the last great respiratory complex, the electrons find their Nirvana (oxygen), but not before another two protons have been shuttled across the membrane. For each pair of electrons stripped from food, ten protons are ferried across the membrane. And that's it (**Figure 9**). A little less than half the energy released by flow of electrons to oxygen is saved in the proton gradient. All that power, all that ingenuity, all the vast protein structures, all of that is dedicated to pumping protons across the inner mitochondrial membrane. One mitochondrion contains tens of thousands of copies of each respiratory complex. A single cell contains hundreds or thousands of mitochondria. Your 40 trillion cells contain at least a quadrillion mitochondria, with a combined convoluted surface area of about 14,000 square *metres*; about four football fields. Their job is to pump protons, and together they pump more than 10^{21} of them – nearly as many as there are stars in the known universe – *every second*.

Well, that's half their job. The other half is to bleed off that power to make ATP.[6] The mitochondrial membrane is very nearly impermeable to protons – that is the point of all these dynamic channels that slam shut as soon as the proton has passed through. Protons are tiny – just the nucleus of the smallest atom, the hydrogen atom – so it is no mean feat to keep them out. Protons pass through water more or less instantaneously, so the membrane must be completely sealed off to water in all places as well. Protons are also charged; they carry a single positive charge. Pumping protons across a sealed membrane achieves two things: first, it generates a difference in the proton concentration between the two sides; and second, it produces a difference in electrical charge, the outside being positive relative to the inside. That means there is an electrochemical potential difference across the membrane, in the

---

6 Not only ATP. The proton gradient is an all-purpose force field, which is used to power the rotation of the bacterial (but not the archaeal) flagellum and the active transport of molecules in and out of the cell, and dissipated to generate heat. It's also central to the life and death of cells by programmed cell death (apoptosis). We'll come to all of that.

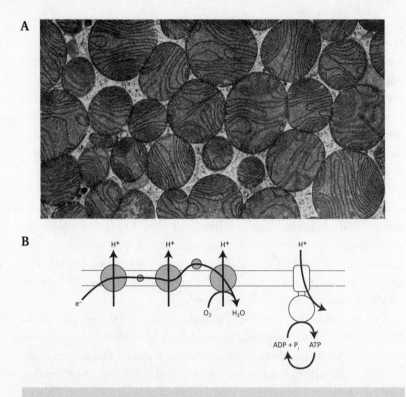

Figure 9 **How mitochondria work**

A Electron micrograph of mitochondria, showing the convoluted inner membranes (cristae) where respiration takes place. [Don Fawcett/Photo Researchers] **B** A cartoon of the respiratory chain, depicting the three major protein complexes embedded in the inner membrane. Electrons (e^-) enter from the left and pass through three large protein complexes to oxygen. The first is complex I (see Figure 8 for a more realistic depiction); electrons then pass through complex III and IV. Complex II (not shown) is a separate entry point into the respiratory chain, and passes electrons straight to complex III. The small circle within the membrane is ubiquinone, which shuttles electrons from complexes I and II to III; the protein loosely bound to the membrane surface is cytochrome *c*, which shuttles electrons from complex III to IV. The current of electrons to oxygen is depicted by the arrow. This current powers the extrusion of protons (H^+) through the three respiratory complexes (complex II passes on electrons but does not pump protons). For each pair of electrons that passes down the chain, four protons are pumped at complex I, four at complex III and two at complex IV. The flux of protons back through the ATP synthase (shown at the right) drives the synthesis of ATP from ADP and P_i.

order of 150 to 200 millivolts. Because the membrane is very thin (around 6 nm thick) this charge is extremely intense across a short distance. Shrink yourself back down to the size of an ATP molecule again, and the intensity of the electric field you would experience in the vicinity of the membrane – the field strength – is 30 million volts per metre, equal to a bolt of lightning, or a thousand times the capacity of normal household wiring.

This huge electrical potential, known as the proton-motive force, drives the most impressive protein nanomachine of them all, the ATP synthase (**Figure 10**). Motive implies motion and the ATP synthase is indeed a rotary motor, in which the flow of protons turns a crank shaft, which in turn rotates a catalytic head. These mechanical forces drive the synthesis of ATP. The protein works like a hydroelectric turbine, whereby protons, pent up in a reservoir behind the barrier of the membrane, flood through the turbine like water cascading downhill, turning the rotating motor. This is barely poetic licence but a precise description, yet it is hard to convey the astonishing complexity of this protein motor. We still don't know exactly how it works – how each proton binds on to the C-ring within the membrane, how electrostatic interactions spin this ring in one direction only, how the spinning ring twists the crank shaft, forcing conformational changes in the catalytic head, how the clefts that open and close in this head clasp ADP and P_i and force them together in mechanical union, to press a new ATP. This is precision nanoengineering of the highest order, a magical device, and the more we learn about it the more marvellous it becomes. Some see in it proof for the existence of God. I don't. I see the wonder of natural selection. But it is undoubtedly a wondrous machine.

For every ten protons that pass through the ATP synthase, the rotating head makes one complete turn, and three newly minted ATP molecules are released into the matrix. The head can spin at over a hundred revolutions per second. I mentioned that ATP is called the universal energy 'currency' of life. The ATP synthase and the proton-motive force are also conserved universally across life. And I mean universally. The ATP synthase is found in basically all bacteria, all archaea, and all eukaryotes (the three domains of life we discussed in the previous chapter), barring a handful of bugs that rely on fermentation instead. It is as universal as the genetic code itself. In

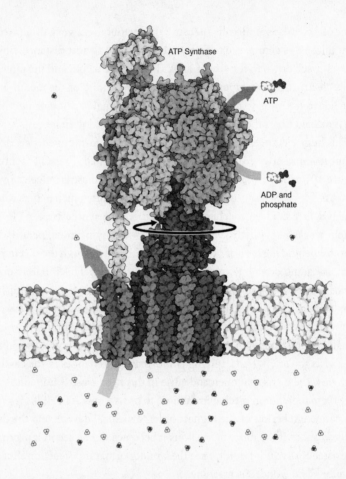

Figure 10 **Structure of the ATP synthase**

The ATP synthase is a remarkable rotating motor embedded in the membrane (bottom). This beautiful artistic rendition by David Goodsell is to scale, and shows the size of an ATP and even protons relative to the membrane and the protein itself. The flux of protons through a membrane subunit (open arrow) drives the rotation of the striped F_O motor in the membrane, as well as the drive shaft (stalk) attached above (turning black arrow). The rotation of the drive shaft forces conformational changes in the catalytic head (F_1 subunit), driving the synthesis of ATP from ADP and phosphate. The head itself is prevented from rotating by the 'stator' – the rigid stick to the left – which fixes the catalytic head in position. Protons are shown below the membrane bound to water as hydronium ions (H_3O^+).

my book, the ATP synthase should be as symbolic of life as the double helix of DNA. Now that you mention it, this is my book, and it is.

A central puzzle in biology

The concept of the proton-motive force came from one of the most quietly revolutionary scientists of the twentieth century, Peter Mitchell. Quiet only because his discipline, bioenergetics, was (and still is) something of a backwater in a research world entranced by DNA. That fascination began in the early 1950s with Crick and Watson in Cambridge, where Mitchell was an exact contemporary. Mitchell, too, went on to win the Nobel Prize, in 1978, but his ideas were far more traumatic in the making. Unlike the double helix, which Watson immediately declared to be 'so pretty it has to be true' – and he was right – Mitchell's ideas were extremely counterintuitive. Mitchell himself was irascible, argumentative and brilliant by turns. He was obliged to retire with stomach ulcers from Edinburgh University in the early 1960s, soon after introducing his 'chemiosmotic hypothesis' in 1961 (which was published in *Nature*, like Crick and Watson's more famous earlier treatise). 'Chemiosmotic' is the term Mitchell used to refer to the transfer of protons across a membrane. Characteristically, he used the word 'osmotic' in its original Greek sense, meaning 'to push' (not in the more familiar usage of osmosis, the passage of water across a semipermeable membrane). Respiration pushes protons across a thin membrane, against a concentration gradient, and hence is chemiosmotic.

With private means and a practical bent, Mitchell spent two years refurbishing a manor house near Bodmin in Cornwall as a lab and home, and opened the Glynn Institute there in 1965. For the next two decades, he and a small number of other leading figures in bioenergetics set about testing the chemiosmotic hypothesis to destruction. The relationships between them took a similar battering. This period has gone down in the annals of biochemistry as the 'ox phos wars' – 'ox phos' being short for 'oxidative phosphorylation', the mechanism by which the flow of electrons to oxygen is coupled to the synthesis of ATP. It's hard to appreciate that none of the details I have given in the last few pages were known as recently as the

1970s. Many of them are still the focus of active research.[7]

Why were Mitchell's ideas so hard to accept? In part because they were so genuinely unexpected. The structure of DNA makes perfect sense – the two strands each act as a template for the other, and the sequence of letters encodes the sequence of amino acids in a protein. The chemiosmotic hypothesis, in comparison, seemed quirky in the extreme, and Mitchell himself might as well have been talking Martian. Life is about chemistry, we all know that. ATP is formed from the reaction of ADP and phosphate, so all that was needed was the transfer of one phosphate from some reactive intermediate on to ADP. Cells are filled with reactive intermediates, so it was just a case of finding the right one. Or so it seemed for several decades. Then along came Mitchell with a mad glint in his eye, plainly an obsessive, writing out equations that nobody could understand, and declaring that respiration was not about chemistry at all, that the reactive intermediate which everyone had been searching for did not even exist, and that the mechanism coupling electron flow to ATP synthesis was actually a gradient of protons across an impermeable membrane, the proton-motive force. No wonder he made people cross!

This is the stuff of legend: a nice example of how science works in unexpected ways, touted as a 'paradigm shift' in biology supporting Thomas Kuhn's view of scientific revolutions, but now safely confined to the history books. The details have been worked out at atomic resolution, culminating in John Walker's Nobel Prize in 1997 for the structure of the ATP synthase. Resolving the structure of complex I is an even taller order, but outsiders might be forgiven for thinking that these are details, and that bioenergetics is no longer hiding any revolutionary discoveries to compare with Mitchell's own. That's ironic because Mitchell arrived at his radical view of bioenergetics not by thinking about the detailed mechanism of respiration

---

7 I am privileged that my office is down the corridor from Peter Rich, who headed the Glynn Institute after Peter Mitchell's retirement, and finally brought it to UCL as the Glynn Laboratory of Bioenergetics. He and his group are working actively on the dynamic water channels that conduct protons through complex IV (cytochrome oxidase), the final respiratory complex in which oxygen is reduced to water.

itself, but a much simpler and more profound question – how do cells (he had bacteria in mind) keep their insides different from the outside? From the very beginning, he saw organisms and their environment as intimately and inextricably linked through membranes, a view which is central to this whole book. He appreciated the importance of these processes to the origin and existence of life in a way that very few others have done since. Consider this passage from a lecture that he gave on the origin of life in 1957, at a meeting in Moscow, four years before publishing his chemiosmotic hypothesis:

> I cannot consider the organism without its environment... From a formal point of view the two may be regarded as equivalent phases between which dynamic contact is maintained by the membranes that separate and link them.

This line of Mitchell's thinking is more philosophical than the nuts and bolts of the chemiosmotic hypothesis, which grew from it, but I think it is equally prescient. Our modern focus on molecular biology means we have all but forgotten Mitchell's preoccupation with membranes as a necessary link between inside and outside, with what Mitchell called 'vectorial chemistry' – chemistry with a direction in space, where position and structure matter. Not test-tube chemistry, where everything is mixed in solution. Essentially all life uses redox chemistry to generate a gradient of protons across a membrane. Why on earth do we do that? If these ideas seem less outrageous now than they did in the 1960s, that is only because we have lived with them for 50 years, and familiarity breeds, if not contempt, at least dwindling interest. They have collected dust and settled into textbooks, ne'er to be questioned again. We now know that these ideas are true; but are we any closer to knowing why they are true? The question boils down to two parts: why do all living cells use redox chemistry as a source of free energy? And why do all cells conserve this energy in the form of proton gradients over membranes? At a more fundamental level, these questions are: why electrons, and why protons?

Life is all about electrons

So why does life on earth use redox chemistry? Perhaps this is the easiest part to answer. Life as we know it is based on carbon, and specifically on partially reduced forms of carbon. To an absurd first approximation (putting aside the requirement for relatively small amounts of nitrogen, phosphorus, and other elements), a 'formula' for life is CH_2O. Given the starting point of carbon dioxide (more on this in the next chapter), then life must involve the transfer of electrons and protons from something like hydrogen (H_2) on to CO_2. It doesn't matter in principle where those electrons come from – they could be snatched from water (H_2O) or hydrogen sulphide (H_2S) or even ferrous iron (Fe^{2+}). The point is they are transferred on to CO_2, and all such transfers are redox chemistry. 'Partially reduced', incidentally, means that CO_2 is not reduced completely to methane (CH_4).

Could life have used something other than carbon? No doubt it is conceivable. We are familiar with robots made from metal or silicon, so what is special about carbon? Quite a lot, in fact. Each carbon atom can form four strong bonds, much stronger than the bonds formed by its chemical neighbour silicon. These bonds allow an extraordinary variety of long-chain molecules, notably proteins, lipids, sugars and DNA. Silicon can't manage anything like this wealth of chemistry. What's more, there are no gaseous silicon oxides to compare with carbon dioxide. I imagine CO_2 as a kind of a Lego brick. It can be plucked from the air and added one carbon at a time on to other molecules. Silicon oxides in contrast ... well, you try building with sand. Silicon or other elements might be amenable to use by a higher intelligence such as ourselves, but it is hard to see how life could have bootstrapped itself from the bottom up using silicon. That's not to say that silicon-based life couldn't possibly evolve in an infinite universe, who could say; but as a matter of probability and predictability, which is what this book is about, that seems overwhelmingly less likely. Apart from being much better, carbon is also much more abundant across the universe. To a first approximation, then, life should be carbon based.

But the requirement for partially reduced carbon is only a small part of the answer. In most modern organisms, carbon metabolism is quite separate from energy metabolism. The two are connected by ATP and a handful of other reactive intermediates such as thioesters (notably acetyl CoA) but there is no fundamental requirement for these reactive intermediates to be produced by redox chemistry. A few organisms survive by fermentation, though this is neither ancient nor impressive in yield. But there is no shortage of ingenious suggestions about possible chemical starting points for life, one of the most popular (and perverse) being cyanide, which could be formed by the action of UV radiation on gases such as nitrogen and methane. Is that feasible? I mentioned in the last chapter that there is no hint from zircons that the early atmosphere contained much methane. That doesn't mean it couldn't happen in principle on another planet, though. And if it's possible, why shouldn't it power life today? We'll return to this in the next chapter. I think it's unlikely for other reasons.

Consider the problem the other way around: what is good about the redox chemistry of respiration? Plenty, it seems. When I say respiration, we need to look beyond ourselves. We strip electrons from food and transfer them down our respiratory chains to oxygen, but the critical point here is that the source and the sink of electrons can both be changed. It so happens that burning up food in oxygen is as good as it gets in terms of energy yield, but the underlying principle is enormously wider and more versatile. There is no need to eat organic matter, for example. Hydrogen gas, hydrogen sulphide and ferrous iron are all electron donors, as we've already noted. They can pass their electrons into a respiratory chain, so long as the acceptor at the other end is a powerful enough oxidant to pull them through. That means bacteria can 'eat' rocks or minerals or gases, using basically the same protein equipment that we use in respiration. Next time you see a discoloration in a concrete wall, betraying a thriving bacterial colony, consider for a moment that, however alien they may seem, they are living by using the same basic apparatus as you.

There's no requirement for oxygen either. Lots of other oxidants can do the job nearly as well, such as nitrate or nitrite, sulphate or sulphite. The list goes on and on. All these oxidants (so called because they behave a little like oxygen) can suck up electrons from food or other sources. In each case, the transfer of electrons from an electron donor to an acceptor releases energy that is stored in the bonds of ATP. An inventory of all known electron donors and electron acceptors used by bacteria and archaea – so-called 'redox couples' – would extend over several pages. Not only do bacteria 'eat' rocks, but they can 'breathe' them too. Eukaryotic cells are pathetic in comparison. There is about the same metabolic versatility in the entire eukaryotic domain – all plants, animals, algae, fungi and protists – as there is in a single bacterial cell.

This versatility in the use of electron donors and acceptors is aided by the sluggish reactivity of many of them. We noted earlier that all biochemistry occurs spontaneously, and must always be driven by a highly reactive environment; but if the environment is too reactive, then it will go right ahead and react, and there will be no free energy left over to power biology. An atmosphere could never be full of fluorine gas, for example, as it would immediately react with everything and disappear. But many substances accumulate to levels that far exceed their natural thermodynamic equilibrium, because they react very slowly. If given a chance, oxygen will react vigorously with organic matter, burning everything on the planet, but this propensity to violence is tempered by a lucky chemical quirk that makes it stable over aeons. Gases such as methane and hydrogen will react even more vigorously with oxygen – just think of the Hindenburg airship – but again, the kinetic barrier to their reaction means that all these gases can coexist in the air for years at a time, in dynamic disequilibrium. The same applies to many other substances, from hydrogen sulphide to nitrate. They can be coerced into reacting, and when they do they release a large amount of energy that can be harnessed by living cells; but without the right catalysts, nothing much happens. Life exploits these kinetic barriers, and in so doing increases entropy faster than would otherwise happen. Some even define life in these terms, as an entropy generator. Regardless: life exists precisely because kinetic barriers exist – it specialises to break them down. Without the loophole of great reactivity pent up behind kinetic barriers, it's doubtful that life could exist at all.

The fact that many electron donors and acceptors are both soluble and stable, entering and exiting cells without much ado, means that the reactive environment required by thermodynamics can be brought safely inside, right into those critical membranes. That makes redox chemistry much easier to deal with than heat or mechanical energy, or UV radiation or light-ning, as a form of biologically useful energy flux. Health and Safety would approve.

Perhaps unexpectedly, respiration is also the basis of photosynthesis. Recall that there are several types of photosynthesis. In each case, the energy of sunlight (as photons) is absorbed by a pigment (usually chloro-phyll) which excites an electron, sending it off down a chain of redox centres to an acceptor, in this case carbon dioxide itself. The pigment, bereft of an electron, accepts one gratefully from the nearest donor, which could be water, hydrogen sulphide, or ferrous iron. As in respiration, the identity of the electron donor doesn't matter in principle. 'Anoxygenic' forms of photosynthesis use hydrogen sulphide or iron as electron donors, leaving behind brimstone or rusty iron deposits as waste.[8] Oxygenic photosynthesis uses a much tougher donor, water, releasing oxygen gas as waste. But the point is that all of these different types of photosynthesis obviously derive from respiration. They use exactly the same respiratory proteins, the same types of redox centre, the same proton gradients over membranes, the same ATP synthase – all the same kit.[9] The only real difference is the innovation of a pigment, chlorophyll, which is in any case closely related to the pigment haem, used in many ancient respiratory proteins. Tapping into the energy of the sun changed the world, but in molecular terms all it did was set elec-trons flowing faster down respiratory chains.

The great advantage of respiration, then, is its immense versatility. Essentially any redox couple (any pair of electron donor and electron acceptor) can be used to set electrons flowing down respiratory chains. The

---

8 This is one of the disadvantages of anoxygenic photosynthesis – cells ultimately encase themselves in their own waste product. Some banded-iron formations are pitted with tiny bacteria-sized holes, presumably reflecting just that. In contrast, oxygen, though potentially toxic, is a much better waste product as it is a gas that simply diffuses away.
9 How can we be so sure it was that way round, rather than respiration deriving from photosynthesis? Because respiration is universal across all life, but photosynthesis is restricted to just a few groups of bacteria. If the last universal common ancestor were photosynthetic, then most groups of bacteria and *all* archaea must have lost this valuable trait. That's not parsimonious, to say the least.

specific proteins that pick up electrons from ammonium are slightly different from those that pick up electrons from hydrogen sulphide, but they are very closely related variations on a theme. Likewise, at the other end of the respiratory chain, the proteins that pass electrons on to nitrate or nitrite differ from those that pass electrons on to oxygen, but all of them are related. They're sufficiently similar to each other that one can be used in place of another. Because these proteins are plugged into a common operating system, they can be mixed and matched to fit any environment. They are not only interchangeable in principle, but in practice they're passed around with abandon. Over the past few decades, we've come to realise that lateral gene transfer (passing around little cassettes of genes from one cell to another, as if spare change) is rife in bacteria and archaea. Genes encoding respiratory proteins are among those most commonly swapped by lateral transfer. Together, they comprise what biochemist Wolfgang Nitschke calls the 'redox protein construction kit'. Did you just move to an environment where hydrogen sulphide and oxygen are both common, such as a deep-sea vent? No problem, help yourself to the requisite genes, they'll work just fine for you, sir. You've run out of oxygen? Try nitrite, ma'am! Don't worry. Take a copy of nitrite reductase and plug it in, you'll be fine!

All these factors mean that redox chemistry should be important for life elsewhere in the universe too. While we could imagine other forms of power, the requirement for redox chemistry to reduce carbon, combined with the many advantages of respiration, means it is hardly surprising that life on earth is redox powered. But the actual mechanism of respiration, proton gradients over membranes, is another matter altogether. The fact that respiratory proteins can be passed round by lateral gene transfer, and mixed and matched to work in any environment, is largely down to the fact that there is a common operating system – chemiosmotic coupling. Yet there is no obvious reason why redox chemistry should involve proton gradients. That lack of an intelligible connection partly explains the resistance to Mitchell's ideas, and the ox phos wars, all those years ago. Over the past 50 years, we have learned a lot about *how* life uses protons; but until we know *why* life uses protons, we will not be able to predict much else about the properties of life here or anywhere else in the universe.

Life is all about protons

The evolution of chemiosmotic coupling is a mystery. The fact that all life is chemiosmotic implies that chemiosmotic coupling arose very early indeed in evolution. Had it arisen later on, it would be difficult to explain how and why it became universal – why proton gradients displaced everything else completely. Such universality is surprisingly rare. All life shares the genetic code (again, with a few minor exceptions, which prove the rule). Some fundamental informational processes are also universally conserved. For example, DNA is transcribed into RNA, which is physically translated into proteins on nanomachines called ribosomes in all living cells. But the differences between archaea and bacteria are really shocking. Recall that the bacteria and archaea are the two great domains of prokaryotes, cells that lack a nucleus and indeed most of the paraphernalia of complex (eukaryotic) cells. In their physical appearance, bacteria and archaea are virtually indistinguishable; but in much of their biochemistry and genetics, the two domains are radically different.

Take DNA replication, which we might guess would be as fundamental to life as the genetic code. Yet the detailed mechanisms of DNA replication, including almost all the enzymes needed, turn out to be totally different in bacteria and archaea. Likewise, the cell wall, the rigid outer layer that protects the flimsy cell inside, is chemically completely different in bacteria and archaea. So are the biochemical pathways of fermentation. Even the cell membranes – strictly necessary for chemiosmotic coupling, otherwise known as *membrane* bioenergetics – are biochemically different in bacteria and archaea. In other words, the barriers between the inside and outside of cells and the replication of hereditary material are not deeply conserved. What could be more important to the lives of cells than these! In the face of all that divergence, chemiosmotic coupling is universal.

These are profoundly deep differences, and lead to sobering questions about the common ancestor of both groups. Assuming that traits in common were inherited from a shared ancestor, but traits that differ arose independently in the two lines, what manner of a cell could that ancestor have been? It defies logic. At face value it was a phantom of a cell, in some ways like modern cells, in other ways . . . well, what exactly? It had DNA transcription, ribosomal translation, an ATP synthase, bits and pieces of amino acid biosynthesis, but beyond that, little else that is conserved in both groups.

Consider the membrane problem. Membrane bioenergetics are universal – but membranes are not. One might imagine that the last common ancestor had a bacterial-type membrane, and that archaea replaced it for some adaptive reason, perhaps because archaeal membranes are better at higher temperatures. That is superficially plausible, but there are two big problems. First, most archaea are not hyperthermophiles; many more live in temperate conditions, in which archaeal lipids offer no obvious advantage; and conversely plenty of bacteria live happily in hot springs. Their membranes cope perfectly well with high temperatures. Bacteria and archaea live alongside each other in almost all environments, frequently in very close symbioses. Why would one of these groups have gone to the serious trouble of replacing all their membrane lipids, on just one occasion? If it is possible to switch membranes, then why don't we see the wholesale replacement of membrane lipids on other occasions, as cells adapt to new environments? That should be much easier than inventing new ones from scratch. Why don't some bacteria living in hot springs acquire archaeal lipids?

Second, and more telling, a major distinction between bacterial and archaeal membranes seems to be purely random – bacteria use one stereo-isomer (mirror form) of glycerol, while archaea use the other.[10] Even if archaea really did replace all their lipids because they were better adapted to high temperatures, there is no conceivable selective reason to replace

---

10 Lipids are composed of two parts: a hydrophilic head-group, and two or three hydrophobic 'tails' (fatty acids in bacteria and eukaryotes, and isoprenes in archaea). These two parts enable lipids to form bilayers, rather than fatty droplets. The head-group in archaea and bacteria is the same molecule, glycerol, but they each use the opposite mirror image form. This is an interesting tangent to the commonly cited fact that all life uses left-handed amino acids and right-handed sugars in DNA. This chirality is often explained in terms of some sort of abiotic prejudice for one isomer over the other, rather than selection at the level of biological enzymes. The fact that archaea and bacteria use the opposite stereoisomers of glycerol shows that chance and selection probably played a large role.

glycerol with glycerol. That's just perverse. Yet the enzyme that makes the left-handed form of glycerol is not even remotely related to the enzyme that makes the right-handed type. To switch from one isomer to the other would require the 'invention' of a new enzyme (to make the new isomer) followed by the systematic elimination of the old (but fully functional) enzyme in each and every cell, even though the new version offered no evolutionary advantage. I just don't buy that. But if one type of lipid was not physically replaced with another, then what kind of membrane did the last common ancestor actually possess? It must have been very different from all modern membranes. Why?

There are also challenging problems with the idea that chemiosmotic coupling arose very early in evolution. One is the sheer sophistication of the mechanism. We have already paid our dues to the giant respiratory complexes and the ATP synthase – incredible molecular machines with pistons and rotary motors. Could these really be a product of the earliest days of evolution, before the advent of DNA replication? Surely not! But that's a purely emotional response. The ATP synthase is no more complex than a ribosome, and everyone is agreed that ribosomes had to evolve early. The second problem is the membrane itself. Even putting aside the question of what type of membrane it was, there is again the issue of disturbing early sophistication. In modern cells, chemiosmotic coupling only works if the membrane is almost impermeable to protons. But all experiments with plausible early membranes suggest that they would have been highly permeable to protons. It's extremely difficult to keep them out. The problem is that chemiosmotic coupling looks to be useless until a number of sophisticated proteins have been embedded in a proton-tight membrane; and then, but only then, does it serve a purpose. So how on earth did all the parts evolve in advance? It's a classic chicken and egg problem. What's the point of learning to pump protons if you have no way to tap the gradient? And what's the point of learning to tap a gradient, if you have no way of generating one? I'll put forward a possible resolution in Chapter 4.

I closed the first chapter with some big questions about the evolution of life on earth. Why did life arise so early? Why did it stagnate in morphological complexity for several billion years? Why did complex, eukaryotic, cells arise just once in 4 billion years? Why do all eukaryotes share a number of perplexing traits that are never found in bacteria or archaea, from sex and two sexes to ageing? Here I am adding two more questions of an equally unsettling magnitude: why does all life conserve energy in the form of proton gradients across membranes? And how (and when) did this peculiar but fundamental process evolve?

I think the two sets of questions are linked. In this book, I will argue that natural proton gradients drove the origin of life on earth in a very particular environment, but an environment that is almost certainly ubiquitous across the cosmos: the shopping list is just rock, water and CO_2. I will argue that chemiosmotic coupling constrained the evolution of life on earth to the complexity of bacteria and archaea for billions of years. A singular event, in which one bacterium somehow got inside another one, overcame these endless energetic constraints on bacteria. That endosymbiosis gave rise to eukaryotes with genomes that swelled over orders of magnitude, the raw material for morphological complexity. The intimate relationship between the host cell and its endosymbionts (which went on to become mitochondria) was, I shall argue, behind many strange properties shared by eukaryotes. Evolution should tend to play out along similar lines, guided by similar constraints, elsewhere in the universe. If I am right (and I don't for a moment think I will be in all the details, but I hope that the bigger picture is correct) then these are the beginnings of a more predictive biology. One day it may be possible to predict the properties of life anywhere in the universe from the chemical composition of the cosmos.

PART II

THE ORIGIN OF LIFE

3

ENERGY AT LIFE'S ORIGIN

M edieval watermills and modern hydroelectric power stations are
powered by the channelling of water. Funnel the flow into a confined
channel and its force increases. Now it can drive work such as turning a
waterwheel. Conversely, allow the flow to spread out across a wider basin,
and the force diminishes. In a river, it becomes a pond or a ford. You might
attempt to make a crossing, safe in the knowledge that you are unlikely to
be swept away by the force of the current.

Living cells work in a similar way. A metabolic pathway is like a water
channel, except that the flow is of organic carbon. In a metabolic pathway,
a linear sequence of reactions is catalysed by a series of enzymes, each one
acting on the product of the previous enzyme. This constrains the flow of
organic carbon. A molecule enters a pathway, undergoes a succession of
chemical modifications, and exits as a different molecule. The succession
of reactions can be repeated reliably, with the same precursor entering and
the same product leaving each time. With their various metabolic pathways,
cells are like networks of water mills, in which the flow is always confined
within interconnecting channels, always maximised. Such ingenious channelling means that cells need far less carbon and energy to grow than they
would if flow were unconstrained. Rather than dissipating the force at each
step – molecules 'escaping' to react with something else – enzymes keep
biochemistry on the straight and narrow. Cells don't need a great river

surging to the sea, but drive their mills using smaller channels. From an energetic point of view, the power of enzymes is not so much that they speed up reactions, but that they channel their force, maximising the output.

So what happened at the origin of life, before there were any enzymes? Flow was necessarily less constrained. To grow – to make more organic molecules, to double, ultimately to replicate – must have cost more energy, more carbon, not less. Modern cells minimise their energy requirements, but we have already seen that they still get through colossal amounts of ATP, the standard energy 'currency'. Even the simplest cells, which grow from the reaction of hydrogen with carbon dioxide, produce about 40 times as much waste from respiration as new biomass. In other words, for every gram of new biomass produced, the energy-releasing reactions that support this production must generate at least 40 grams of waste. Life is a side reaction of a main energy-releasing reaction. That remains the case today, after 4 billion years of evolutionary refinement. If modern cells produce 40 times more waste than organic matter, just think how much the first primitive cells, without any enzymes, would have had to make! Enzymes speed up chemical reactions by millions of times the unconstrained rate. Take away those enzymes, and throughput would need to increase by a similar factor, say a millionfold, to achieve the same thing. The first cells may have needed to produce 40 tonnes of waste – literally a truck-load – to make 1 gram of cell! In terms of energy flow, that dwarfs even a river in spate; it's more like a tsunami.

The sheer scale of this energetic demand has connotations for all aspects of the origin of life, yet is rarely considered explicitly. As an experimental discipline, the origin-of-life field dates back to 1953 and the famous Miller–Urey experiment, published in the same year as Watson and Crick's double-helix paper. Both papers have hung over the field ever since, casting a shadow like the wings of two giant bats, in some respects rightly, in others regrettably. The Miller–Urey experiment, brilliant as it was, bolstered the conception of a primordial soup, which in my view has blinkered the field for two generations. Crick and Watson ushered in the hegemony of DNA and information, which is plainly of vital importance to the origin of life; but considering replication and the origins of natural selection in near isolation has distracted attention from the importance of other factors, notably energy.

In 1953, Stanley Miller was an earnest young PhD student in the lab of Nobel laureate Harold Urey. In his iconic experiment, Miller passed electrical discharges, simulating lightning, through flasks containing water and a mixture of reduced (electron-rich) gases reminiscent of the atmosphere of Jupiter. At the time, the Jovian atmosphere was thought to reflect that of the early earth – both were presumed to be rich in hydrogen, methane and ammonia.[1] Amazingly, Miller succeeded in synthesising a number of amino acids, which are the building blocks of proteins, the workhorses of cells. Suddenly the origin of life looked easy! In the early 1950s there was far more interest in this experiment than in Watson and Crick's structure, which initially caused little stir. Miller, in contrast, featured on the cover of *Time* magazine in 1953. His work was seminal, still worth recapitulating, because it was the first to test an explicit hypothesis about the origin of life: that bolts of lightning, passing through an atmosphere of reduced gases, could produce the building blocks of cells. In the absence of existing life, these precursors were taken to accumulate in the oceans, which over time became a rich broth of organic molecules, the primordial soup.

If Watson and Crick made less of a stir in 1953, the spell of DNA has beguiled biologists ever since. For many people, life is all about the information copied in DNA. The origin of life, for them, is the origin of information, without which, all are agreed, evolution by natural selection is not possible. And the origin of information boils down to the origin of replication: how the first molecules that made copies of themselves – replicators – arose. DNA itself is too complex to be credible as the first replicator, but the simpler, more reactive precursor RNA fits the bill. RNA (ribonucleic acid) is, even today, the key intermediary between DNA and proteins, serving as both a template and a catalyst in protein synthesis. Because RNA can act as both a template (like DNA) and a catalyst (like proteins), it can in

---

1 Based on the chemistry of zircon crystals and the earliest rocks, the early earth is now believed to have had a relatively neutral atmosphere, reflecting volcanic degassing, and composed mostly of carbon dioxide, nitrogen, and water vapour.

principle serve as a simpler forerunner of both proteins and DNA in a primordial 'RNA world'. But where did all the nucleotide building blocks come from, which join together into chains to form RNA? The primordial soup, of course! There is no necessary relationship between the formation of RNA and a soup, but soup is nonetheless the simplest assumption, which avoids worrying about complicated details like thermodynamics or geochemistry. Put all that to one side, and the gene-jocks can get on with the important stuff. And so, if there has been a leitmotif dominating origin-of-life research over the last 60 years, it is that a primordial soup gave rise to an RNA world, in which these simple replicators gradually evolved and became more complex, began coding for metabolism, and ultimately spawned the world of DNA, proteins and cells that we know today. By this view, life is information from the bottom up.

What is missing here is energy. Of course, energy figures in the primordial soup – all those flashes of lightning. I once calculated that to sustain a tiny primitive biosphere, equivalent in size to that before the evolution of photosynthesis, by lightning alone, would require four bolts of lightning per second, for every square kilometre of ocean. And that's assuming a modern efficiency of growth. There are just not all that many electrons in each bolt of lightning. A better alternative source of energy is UV radiation, which can fashion reactive precursors like cyanide (and derivatives like cyanamide) from a mixture of atmospheric gases including methane and nitrogen. UV radiation streams in endlessly on the earth and other planets. UV flux would have been stronger in the absence of an ozone layer, and with the more aggressive electromagnetic spectrum of the young sun. The ingenious organic chemist John Sutherland has even succeeded in synthesising activated nucleotides under so-called 'plausible primordial conditions' using UV radiation and cyanide.[2] But there are serious problems here too.

---

2 That innocuous phrase 'plausible primordial conditions' actually conceals a multitude of sins. On the face of it, it means simply that the compounds and conditions used could reasonably have been found on the early earth. It is indeed plausible that there was some cyanide in the Hadean oceans; also that temperatures could range between several hundred degrees (in hydrothermal vents) and freezing on the early earth. The trouble

No life on earth uses cyanide as a source of carbon; and no known life uses UV radiation as a source of energy. Quite the contrary, both are considered dangerous killers. UV is too destructive, even for the sophisticated life forms of today, as it breaks down organic molecules rather more effectively than it promotes their formation. It is much more likely to scorch the oceans than to fill them with life. UV is a blitz. I doubt it would work as a direct source of energy, here or anywhere else.

The advocates of UV radiation don't claim that it would work as a direct source of power, but rather that it would favour the formation of small stable organic molecules like cyanide, which accumulate over time. In terms of chemistry, cyanide is indeed a good organic precursor. It is toxic to us because it blocks cell respiration; but that might be a quirk of life on earth, rather than any deeper principle. The real problem with cyanide is its concentration, which afflicts the whole idea of primordial soup. The oceans are extremely large relative to the rate of formation of cyanide, or for that matter of any other simple organic precursor, even assuming that a suitably reducing atmosphere existed here or on any other planet. At any reasonable rate of formation, the steady-state concentration of cyanide in the oceans at 25°C would have been around two-millionths of a gram per litre – not nearly enough to drive the origins of biochemistry. The only way out of this impasse is to concentrate the seawater somehow, and this has been the mainstay of prebiotic chemistry for a generation. Either freezing or evaporating to dryness could potentially increase the concentration of organics, but these are drastic methods, hardly congruent with the physically stable state that is a defining feature of all living cells. One exponent of cyanide origins turns with wild eyes to the great asteroid bombardment 4 billion years ago: it could have concentrated cyanide (as ferricyanide) by evaporat-

---

is that realistic concentrations of organics in a soup are far lower than tend to be used in the lab; and it is hardly feasible to have both heating and freezing in one and the same environment. So yes: all these conditions may have existed somewhere on the planet, but they could only drive prebiotic chemistry if the whole planet is taken as a single unit, engaged in a coherent set of experiments as if it were a synthetic chemist's lab. That is implausible in the extreme.

ing all the oceans! To me, that smacks of desperation to defend an unworkable idea.[3] The problem here is that these environments are too variable and unstable. A succession of drastic changes in conditions are required to achieve the steps to life. In contrast, living cells are stable entities – their fabric is continually replaced, but the overall structure is unchanging.

Heraclitus taught that 'no man ever steps in the same river twice'; but he didn't mean that the river had evaporated or frozen (or been exploded into space) in the meantime. As water flows between unchanging banks, at least on our human timescale, so life is continuously renewing itself without changing its form. Living cells remain cells, even when all their constituent parts are replaced in an unceasing turnover. Could it be any other way? I doubt it. In the absence of information specifying structure – as must logically have been the case at the origin of life, before the advent of replicators – structure is not absent, but it does require a continuous flux of energy. Energy flux promotes self-organisation of matter. We are all familiar with what the great Russian-born Belgian physicist Ilya Prigogine called 'dissipative structures': just think of convection currents in a boiling kettle, or for that matter water swirling down a plughole. No information is required – just heat in the case of the kettle and angular momentum for the plug hole. Dissipative structures are produced by the flux of energy and matter. Hurricanes, typhoons and whirlpools are all striking natural examples of dissipative structures. We find them on a vast scale in the oceans and atmosphere too, driven by the differences in energy flux from the sun at the equator

---

3 I have discussed soup as if it was 'made on earth' by lightning or UV radiation. An alternative source of organics is delivery from space by chemical panspermia. There is no doubt that organic molecules are abundant in space and on asteroids; and there certainly was a steady delivery of organics to earth on meteorites. But once here, these organics must have dissolved in the oceans, at best stocking a primordial soup. That means that chemical panspermia is no answer to the origin of life: it suffers from the same intractable problems as soup. The delivery of whole cells, as advocated by Fred Hoyle, Francis Crick, and others, is likewise no solution: it simply pushes the problem somewhere else. We may never be able to say exactly how life originated on earth, but we can explore the principles that must govern the emergence of living cells here or anywhere else. Panspermia fails utterly to address those principles, and so is irrelevant.

relative to the poles. Reliable ocean currents, such as the Gulf Stream, and winds, such as the Roaring Forties or the North Atlantic jet stream, are not specified by information, but are as stable and continuous as the energy flux that sustains them. The Great Red Spot of Jupiter is a huge storm, an anticyclone several times the size of the earth, which has persisted for at least a few hundred years. Just as the convection cells in a kettle persist for as long as the electric current keeps the water boiling and steam evaporating, all these dissipative structures require a continuous flux of energy. In more general terms, they are the visible products of sustained far-from-equilibrium conditions, in which energy flux maintains a structure indefinitely, until at last (after billions of years in the case of stars) equilibrium is attained and the structure finally collapses. The main point is that sustained and predictable physical structures can be produced by energy flux. This has nothing to do with information, but we'll see that it can create environments where the origin of biological information – replication and selection – is favoured.

All living organisms are sustained by far-from-equilibrium conditions in their environment: we, too, are dissipative structures. The continuous reaction of respiration provides the free energy that cells need to fix carbon, to grow, to form reactive intermediates, to join these building blocks together into long-chain polymers such as carbohydrates, RNA, DNA and proteins, and to maintain their low-entropy state by increasing the entropy of the surroundings. In the absence of genes or information, certain cell structures, such as membranes and polypeptides, should form spontaneously, so long as there is a continuous supply of reactive precursors – activated amino acids, nucleotides, fatty acids; so long as there is a continuous flux of energy providing the requisite building blocks. Cell structures are forced into existence by the flux of energy and matter. The parts can be replaced but the structure is stable and will persist for as long as the flux persists. This continuous flux of energy and matter is precisely what is missing from the primordial soup. There is nothing in soup that can drive the formation of the dissipative structures that we call cells, nothing to make these cells grow and divide, and come alive, all in the absence of enzymes that channel and drive metabolism. That sounds like a tall order. Is there really an environment that can drive the formation of the first primitive cells? There most certainly must have been. But before we explore that environment, let's consider exactly what is needed.

How to make a cell

What does it take to make a cell? Six basic properties are shared by all living cells on earth. Without wishing to sound like a textbook, let's just enumerate them. All need:

(i)   a continuous supply of reactive carbon for synthesising new organics;

(ii)   a supply of free energy to drive metabolic biochemistry – the formation of new proteins, DNA, and so on;

(iii) catalysts to speed up and channel these metabolic reactions;

(iv)   excretion of waste, to pay the debt to the second law of thermodynamics and drive chemical reactions in the correct direction;

(v)   compartmentalisation – a cell-like structure that separates the inside from the outside;

(vi)   hereditary material – RNA, DNA or an equivalent, to specify the detailed form and function.

Everything else (the kind of thing you will find in standard mnemonics for life's properties, such as movement or sensitivity) are just nice-to-have added extras from the point of view of bacteria.

It doesn't take much reflection to appreciate that all six factors are profoundly interdependent, and almost certainly needed to be from the very beginning too. A continuous supply of organic carbon is obviously central to growth, replication, … everything. At a simple level, even an 'RNA world' involves the replication of RNA molecules. RNA is a chain of nucleotide building blocks, each one of which is an organic molecule that must have come from somewhere. There is an old rift between origin-of-life researchers about what came first, metabolism or replication. It's a barren debate. Replication is doubling, which consumes building blocks in an exponential fashion. Unless those building blocks are replenished at a similar rate, replication swiftly ceases.

One conceivable escape is to assume that the first replicators were not organic at all, but were clay minerals or some such, as long and ingeniously argued by Graham Cairns-Smith. Yet that solves little, because minerals are too physically clumsy to *encode* anything even approaching an RNA-world level of complexity, although they are valuable catalysts. But if minerals are no use as replicators, then we need to find the shortest and fastest route to get from inorganics to organic molecules that do work as replicators, like RNA. Given that nucleotides have been synthesised from cyanamide, it's pointless to posit unknown and unnecessary intermediates; it's far better to cut straight to the chase, to assume that some early environments on earth could have provided the organic building blocks – activated nucleotides – needed for the beginnings of replication.[4] Even if cyanamide is a poor starting point, the tendency to produce a strikingly similar spectrum of organics under disparate conditions, from electrical discharges in a reducing atmosphere, to cosmic chemistry on asteroids, to high-pressure bomb reactors, suggests that certain molecules, probably including some nucleotides, are favoured by thermodynamics. To a first approximation, then, the formation of organic replicators requires a continuous supply of organic carbon in the same environment. That rules out freezing environments, incidentally – while freezing can concentrate organics between ice crystals, there is no mechanism to replenish the building blocks needed to continue the process.

What about energy? That is also needed in the same environment. Joining individual building blocks (amino acids or nucleotides) together to form

---

4 This is an appeal to Occam's razor, the philosophical basis of all science: assume the simplest natural cause. That answer might turn out not to be correct, but we should not resort to more complex reasoning unless it is shown to be necessary. We might ultimately need to invoke celestial machinations to explain the origin of replication, when all other possibilities have been disproved (though I doubt it); but until then we should not multiply causes. This is simply a way of approaching a problem; but the remarkable success of science shows that it is a very effective approach.

long-chain polymers (proteins or RNA) requires first activating the building blocks. That in turn demands a source of energy – ATP, or something similar. Perhaps very similar. In a waterworld, as was the earth 4 billion years ago, the source of energy needs to be of a rather specific kind: it needs to drive the polymerisation of long-chain molecules. That involves removing one molecule of water for each new bond formed, a dehydration reaction. The problem of dehydrating molecules in solution is a bit like trying to wring out a wet cloth under water. Some prominent researchers have been so distracted by this problem that they have even contended that life must have started on Mars, where there was much less water. Life then hitch-hiked to earth on a meteorite, making us all Martians really. But of course life here on earth does perfectly well in water. Every living cell pulls off the dehydration trick thousands of times a second. We do so by coupling the dehydration reaction to the splitting of ATP, which takes up one molecule of water each time it is split. Coupling a dehydration to a 'rehydration' reaction (technically termed 'hydrolysis') in effect just transfers the water, while at the same time releasing some of the energy pent up in the bonds of ATP. That greatly simplifies the problem; all that is needed is a continuous supply of ATP or a simpler equivalent, such as acetyl phosphate. We'll address where this may have come from in the next chapter. For now, the point is that replication in water needs a continuous and liberal supply of both organic carbon and something much like ATP, in the same environment.

That's three out of six factors: replication, carbon and energy. What about compartmentalisation into cells? This is again a matter of concentration. Biological membranes are made of lipids, which are themselves composed of fatty acids or isoprenes (joined to a glycerol head-group, as noted in the previous chapter). When concentrated above a threshold level, fatty acids spontaneously form into cell-like vesicles that can grow and divide if continuously 'fed' with new fatty acids. Here again, we need a continuous supply of both organic carbon and energy to drive the formation of new fatty acids. For fatty acids, or for that matter nucleotides, to accumulate faster than they dissipate, there must be some sort of focusing: a physical funnelling or natural compartmentalisation that increases their concentration locally, enabling them to form larger-scale structures. When such conditions are met, the formation of vesicles is not magic: physically, this is the most stable state – overall entropy increases as a result, as we saw in the previous chapter.

If reactive building blocks are indeed supplied continuously, then simple vesicles will grow and divide spontaneously, as a result of surface-area-to-volume constraints. Imagine a spherical vesicle – a simple 'cell' – enclosing various organic molecules. The vesicle grows by incorporating new materials: lipids in the membrane and other organics inside the cell. Now let's double in size: double the membrane surface area, and double the organic contents. What happens? Doubling the surface area more than doubles the volume, because the surface area increases by the square of the radius, while the volume increases by its cube. But the contents only doubled. Unless the contents increase at a faster rate than the membrane surface area, the vesicle will buckle into a dumb-bell, which is already halfway to forming two new vesicles. In other words, arithmetic growth introduces an instability that leads to division and doubling, rather than simply getting bigger. It's only a matter of time before a growing sphere divides up into smaller bubbles. So a continuous flux of reactive carbon precursors entails not only a primitive cell formation but also a rudimentary form of cell division. Such budding, incidentally, is also how L-form bacteria, which lack a cell wall, divide.

The problem of surface-area-to-volume ratio must set a limit to the size of cells. This is just a matter of the supply of reactants and the removal of waste. Nietzsche once observed that humans will not mistake ourselves for gods so long as we need to defecate. But in fact excretion is a thermo-dynamic necessity, binding even for the godliest. For any reaction to continue in a forward direction, the end product must be removed. This is no more mysterious than the build-up of a crowd at a railway station. If passengers can't get on to a train as fast as new people arrive, there will soon be a blockage. In the case of cells, the rate at which new proteins are formed depends on the rates of delivery of reactive precursors (activated amino acids) and removal of waste (methane, water, CO_2, ethanol – whatever the energy-releasing reaction may be). If these waste products are not physically removed from the cell, they prevent the forward reaction from continuing.

The problem of waste removal is another fundamental difficulty with the idea of a primordial soup, in which reactants and waste marinade together. There is no forward momentum, no driving force for new chemistry.[5] Likewise, the larger a cell becomes, the closer it approximates to soup. Because the volume of a cell rises faster than its surface area, the relative rate at which fresh carbon can be delivered and waste removed across its bounding membrane must fall as the cell gets larger. A cell on the scale of the Atlantic ocean, or even a football, could never work; it is just soup. (You might think than an ostrich egg is as big as a football, but the yolk sac is mostly just a food dump – the developing embryo itself is much smaller.) At the origin of life, natural rates of carbon delivery and waste removal must have dictated a small cell volume. Some sort of physical channelling would also seem to be necessary: a continuous natural flow that delivers precursors and carries away waste.

That leaves us with catalysts. Today, life uses proteins – enzymes – but RNA also has some catalytic capabilities. The trouble here is that RNA is already a sophisticated polymer, as we have seen. It is composed of multiple nucleotide building blocks, each of which must be synthesised and activated to join together into a long chain. Before that happened, RNA could hardly have been the catalyst. Whatever process gave rise to RNA must also have driven the formation of other organic molecules that are easier to make, notably amino acids and fatty acids. Thus any early 'RNA world' must have been 'dirty' – contaminated with many other types of small organic molecules. The idea that RNA somehow invented metabolism by itself is absurd, even if RNA did play a key role in the origins of replication and protein synthesis. So what did catalyse the beginnings of biochemistry? The probable answer is inorganic complexes, such as metal

---

5 A familiar example is the alcohol content of wine, which cannot rise above about 15% by alcoholic fermentation alone. As alcohol builds up, it blocks the forward reaction (fermentation), preventing the formation of any more alcohol. Unless the alcohol is removed, fermentation grinds to a halt: the wine has reached thermodynamic equilibrium (it has become soup). Spirits such as brandy are produced by distilling wine, thereby concentrating the alcohol further; I believe we are the only life form that has perfected distillation.

sulphides (in particular iron, nickel and molybdenum). These are still found as cofactors in several ancient, and universally conserved, proteins. While we tend to think of the protein as the catalyst, in fact the protein only speeds up reactions that happen anyway – the cofactor determines the nature of the reaction. Stripped of their protein context, cofactors are not very effective or specific catalysts, but they are much better than nothing. How effective they are depends, yet again, on the throughput. The first inorganic catalysts just began the channelling of carbon and energy in the direction of organics, but they cut the need for a tsunami back down to a mere river.

And these simple organics (notably amino acids and nucleotides) also have some catalytic activity of their own. In the presence of acetyl phosphate, amino acids can even join together, to form short 'polypeptides' – little strings of amino acids. The stability of such polypeptides depends in part on their interactions with other molecules. Hydrophobic amino acids or polypeptides that associate with fatty acids should persist longer; and charged polypeptides that bind to inorganic clusters such as FeS minerals could also be more stable. Natural associations between short polypeptides and mineral clusters may enhance the catalytic properties of minerals, and could be 'selected' for by simple physical survival. Imagine a mineral catalyst that promotes organic synthesis. Some of the products bind to the mineral catalyst, prolonging their own survival, while at once improving (or at least varying) the catalytic properties of the mineral. Such a system could in principle give rise to richer and more complex organic chemistry.

So how could a cell be built from scratch? There must be a continuously high flux of reactive carbon and usable chemical energy, flowing past rudimentary catalysts that convert a modest proportion of that flux into new organics. This continuous flux must be constrained in some way that enables the accumulation of high concentrations of organics, including fatty acids, amino acids and nucleotides, without compromising the outflow of waste. Such a focusing of flow could be achieved by a natural channelling or compartmentalisation, which has the same effect as the channelling of flow in a water mill – it increases the force of a given flux in the absence of enzymes, so lowering the total amount of carbon and energy required. Only if the synthesis of new organics exceeds their rate of loss into the outside world, enabling their concentration, will they self-assemble into structures such as cell-like vesicles, RNA and proteins.[6]

Plainly this is no more than the beginnings of a cell – necessary, but far from sufficient. But let's put aside the details for now, and focus on just this one point. Without a high flux of carbon and energy that is physically channelled over inorganic catalysts, there is no possibility of evolving cells. I would rate this as a necessity anywhere in the universe: given the requirement for carbon chemistry that we discussed in the last chapter, thermodynamics dictates a continuous flow of carbon and energy over natural catalysts. Discounting special pleading, that rules out almost all environments that have been touted as possible settings for the origin of life: warm ponds (sadly Darwin was wrong on that), primordial soup, microporous pumice stones, beaches, panspermia, you name it. But it does not rule out hydrothermal vents; on the contrary, it rules them in. Hydrothermal vents are exactly the kind of dissipative structures that we seek – continuous flow, far-from-equilibrium electrochemical reactors.

Hydrothermal vents as flow reactors

The Grand Prismatic Spring in Yellowstone National Park reminds me of the Eye of Sauron in its malevolent yellows, oranges and greens. These remarkably vivid colours are the photosynthetic pigments of bacteria that use hydrogen (or hydrogen sulphide) emanating from the volcanic springs as an electron donor. Being photosynthetic, the Yellowstone bacteria give little real insight into the origin of life, but they do give a sense of the primal

---

6 I don't really mean proteins, I mean polypeptides. The sequence of amino acids in a protein is specified by a gene, in DNA. A polypeptide is a string of amino acids joined together by the same type of bond, but is usually much shorter (perhaps just a few amino acids) and their sequence does not need to be specified by a gene. Short polypeptides will form spontaneously from amino acids, in the presence of a chemical 'dehydrating' agent such as pyrophosphate or acetyl phosphate, which are plausible abiotic precursors of ATP.

power of volcanic springs. These are plainly hot spots for bacteria, in otherwise meagre environments. Go back 4 billion years, strip away the surrounding vegetation to the bare rocks, and it's easy to imagine such a primal place as the birthplace of life.

Except that it wasn't. Back then, the earth was a waterworld. Perhaps there were a few terrestrial hot springs on small volcanic islands protruding above tempestuous global oceans, but most vents were submerged beneath the waves in deep-sea hydrothermal systems. The discovery of submarine vents in the late 1970s came as a shock, not because their presence was unsuspected (plumes of warm water had betrayed their presence) but because nobody anticipated the brutal dynamism of 'black smokers', or the overwhelming abundance of life clinging precariously to their sides. The deep ocean floor is mostly a desert, nearly destitute of life. Yet these tottering chimneys, billowing out black smoke as if their lives depended on it, were home to peculiar and hitherto unknown animals – giant tube worms lacking a mouth and anus, clams as big as dinner plates, and eyeless shrimp – all living at a density equivalent to tropical rain forests. This was a seminal moment, not only for biologists and oceanographers, but perhaps even more for those interested in the origin of life, as the microbiologist John Baross was quick to appreciate. Since then, Baross more than anyone has kept attention focused on the extraordinary vigour of chemical disequilibria in vents down in the bible-black ocean depths, far away from the sun.

Yet these vents, too, are misleading. They are not really cut off from the sun. The animals that live here rely on symbiotic relationships with bacteria that oxidise the hydrogen sulphide gas emanating from the smokers. That is the principal source of disequilibrium: hydrogen sulphide (H_2S) is a reduced gas that reacts with oxygen to release energy. Recall the mechanics of respiration from the previous chapter. Bacteria use H_2S as an electron donor for respiration, and oxygen as the electron acceptor, to drive ATP synthesis. But oxygen is a side-product of photosynthesis, and was not present on the early earth, before the evolution of oxygenic photosynthesis. The stunning eruption of life around these black smoker vents is therefore completely, albeit indirectly, dependent on the sun. And that means these vents must have been very different 4 billion years ago.

Take away the oxygen and what is left? Well, black smokers are produced by the direct interactions of seawater with magma at the tectonic spreading centres of mid-ocean ridges or other volcanically active places. Water percolates down through the sea floor to the magma chambers not far below, where it is heated instantaneously to hundreds of degrees, and charged with dissolved metals and sulphides, making the water strongly acidic. As the superheated water blasts back up into the ocean above, bursting with explosive power, it cools abruptly. Tiny particles of iron sulphides such as pyrites (fool's gold) precipitate immediately – this is the black smoke that gives these angry volcanic vents their name. Most of that would have been the same 4 billion years ago, but none of this volcanic fury is available to life. Only the chemical gradients matter; and there's the rub. The chemical boost provided by oxygen would have been missing. Trying to get hydrogen sulphide to react with CO_2 to form organics is much harder, especially at high temperatures. In a succession of groundbreaking papers from the late 1980s onwards, the revolutionary and notoriously irascible German chemist and patent attorney Günter Wächtershäuser redrew the landscape.[7] He proposed in great detail a way of reducing CO_2 to organic molecules on the surface of the mineral iron pyrites, which he termed 'pyrites pulling'. More broadly, Wächtershäuser talked of an 'iron–sulphur world', in which iron–sulphur (FeS) minerals catalysed the formation of organic molecules. Such minerals are typically composed of repeating lattices of ferrous iron (Fe^{2+}) and sulphide (S^{2-}). Little mineral clusters of ferrous iron and sulphide, known as FeS clusters, are still found at the heart of many enzymes today, including those involved in respiration. Their structure is essentially identical to the lattice structure of FeS minerals such as mackinawite and greigite

---

7 Wächtershäuser transformed perceptions about the origin of life. He dismissed primordial soup in no uncertain terms, beginning a prolonged and bitter argument in the journals with Stanley Miller. Here's one broadside from Wächtershäuser, for anyone who thinks that science is in some sense dispassionate: 'The prebiotic broth theory has received devastating criticism for being logically paradoxical, incompatible with thermodynamics, chemically and geochemically implausible, discontinuous with biology and biochemistry, and experimentally refuted.'

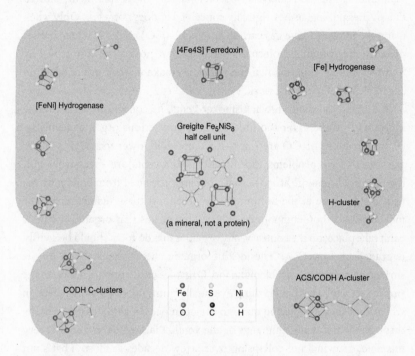

Figure 11 **Iron–sulphur minerals and iron–sulphur clusters**
The close similarity between iron–sulphur minerals and iron–sulphur clusters embedded in modern enzymes, as depicted by Bill Martin and Mike Russell in 2004. The central panel shows a repeating crystalline unit from the mineral greigite; this structure is repeated to make up a lattice of multiple units. The surrounding panels show iron–sulphur clusters embedded in proteins, with structures similar to greigite and related minerals such as mackinawite. The shaded areas represent the rough shape and size of the protein named in each case. Each protein typically contains a few iron–sulphur clusters, with or without nickel.

(**Figure 11**; see also **Figure 8**), lending credence to the idea that these minerals could have catalysed the first steps of life. Nonetheless, even though FeS minerals are good catalysts, Wächtershäuser's own experiments showed that pyrites pulling, as he originally conceived it, does not work. Only when using the more reactive gas carbon monoxide (CO) could Wächtershäuser produce any organic molecules. The fact that no known life grows by 'pyrites pulling' suggests that the failure to make it work in the lab is no accident; it really doesn't work.

While CO is found in black smoker vents, its concentration is vanishingly low – there is far too little to drive any serious organic chemistry. (Concentrations of CO are 1,000–1,000,000-fold lower than CO_2.) There are other grave problems too. Black smoker vents are excessively hot; the vent fluids emerge at 250–400°C, but are prevented from boiling by the extreme pressure at the bottom of the ocean. At these temperatures, the most stable carbon compound is CO_2. This means that organic synthesis can't take place; on the contrary, any organics that do form should be swiftly degraded back to CO_2. The idea of organic chemistry catalysed by the surface of minerals is problematic too. Organics either remain bound to the surface, in which case everything eventually gums up, or they dissociate, in which case they are flushed out into the open oceans with unseemly haste, through the billowing chimneys of the vents. Black smokers are also very unstable, growing and collapsing over a few decades at most. That's not long to 'invent' life. While they are truly far-from-equilibrium dissipative structures, and certainly solve some of the problems of soup, these volcanic systems are too extreme and unstable to nurture the gentle carbon chemistry needed for the origin of life. What they did do, which was indispensable, was charge the early oceans with catalytic metals such as ferrous iron (Fe^{2+}) and nickel (Ni^{2+}) derived from magma.

The beneficiary of all these metals dissolved in the ocean was another type of vent known as alkaline hydrothermal vents (**Figure 12**). In my view, these resolve all the problems of black smokers. Alkaline hydrothermal vents are not volcanic at all, and lack the drama and excitement of black smokers; but they do have other properties that fit them out much better as electrochemical flow reactors. Their relevance to the origin of life was first

Figure 12 **Deep-sea hydrothermal vents**
Comparison of an active alkaline hydrothermal vent at Lost City (**A**) with a black smoker (**B**). The scale bar is one metre in both cases: alkaline vents can stand as much as 60 m tall, equivalent to a 20-storey building. The white arrow at the top marks a probe fixed to the top of the alkaline vent. The whiter regions of alkaline vents are the most active, but unlike black smokers, these hydrothermal fluids do not precipitate as 'smoke'. The sense of abandonment, though misleading, influenced the choice of the name Lost City.

signalled by the revolutionary geochemist Mike Russell, in a short letter to *Nature* in 1988, and developed in a series of idiosyncratic theoretical papers through the 1990s. Later on, Bill Martin brought his inimitable microbiological perspective to bear on the vent world, the pair pointing out many unexpected parallels between vents and living cells. Russell and Martin, like Wächtershäuser, argue that life started from the 'bottom up', through the reaction of simple molecules such as H_2 and CO_2, in much the same way as autotrophic bacteria (which synthesise all their organic molecules from simple inorganic precursors). Russell and Martin likewise always stressed the importance of iron–sulphur (FeS) minerals as early catalysts. The fact that Russell, Martin and Wächtershäuser all talk of hydrothermal vents, FeS minerals and autotrophic origins means that their ideas are easily conflated. In reality, the differences are as black and white.

Alkaline vents are not produced by the interactions of water with magma but by a much gentler process – a chemical reaction between solid rock and water. Rocks derived from the mantle, rich in minerals such as olivine, react with water to become the hydrated mineral serpentinite. This mineral has a beautiful mottled green appearance, which resembles the scales of a serpent. Polished serpentinite is commonly used as an ornamental stone, like green marble, in public buildings, including the United Nations in New York. The chemical reaction that forms the rock has acquired the forbidding name of 'serpentinisation', but all this means is that olivine reacts with water to form serpentinite. The *waste products* of this reaction are key to the origin of life.

Olivine is rich in ferrous iron and magnesium. The ferrous iron is oxidised by water to the rusty ferric oxide form. The reaction is exothermic (releasing heat), and generates a large amount of hydrogen gas, dissolved in warm alkaline fluids containing magnesium hydroxides. Because olivine is common in the earth's mantle, this reaction occurs largely on the sea floor, close to the tectonic spreading centres, where fresh mantle rocks are exposed to ocean waters. Mantle rocks are rarely exposed directly – water percolates down beneath the sea floor, sometimes to depths of several kilometres, where it reacts with olivine. The warm, alkaline, hydrogen-rich fluids produced are more buoyant than the descending cold ocean water, and bubble back up to the sea floor. There they cool, and react with salts dissolved in the ocean, precipitating into large vents on the sea floor.

Unlike black smokers, alkaline vents have nothing to do with magma, and so are not found directly above the magma chambers at the spreading centres, but typically some miles away. They are not superheated, but warm, with temperatures of 60 to 90°C. They are not open chimneys, venting directly into the sea, but riddled with a labyrinth of interconnected micropores. And they are not acidic, but strongly alkaline. Or at least, these are the properties that Russell predicted in the early 1990s on the basis of his theory. His was a lone and impassioned voice at conferences, arguing that scientists were mesmerised by the dramatic vigour of black smokers, and overlooking the quieter virtues of alkaline vents. Not until the discovery of the first known submarine alkaline vent in the year 2000, dubbed Lost City, did researchers really begin to listen. Lost City, remarkably, conforms to almost all of Russell's predictions, right down to its location, some 10 miles from the mid-Atlantic Ridge. As it happens, this was the time that I first began thinking and writing about bioenergetics in relation to the origins of life (my book *Oxygen* was published in 2002). These ideas were immediately appealing: for me, the wonderful reach of Russell's hypothesis is that, uniquely, it ties in natural proton gradients to the origin of life. The question is: how, exactly?

The importance of being alkaline

Alkaline hydrothermal vents provide exactly the conditions required for the origin of life: a high flux of carbon and energy that is physically channelled over inorganic catalysts, and constrained in a way that permits the accumulation of high concentrations of organics. The hydrothermal fluids are rich in dissolved hydrogen, with lesser quantities of other reduced gases including methane, ammonia and sulphide. Lost City and other known alkaline vents are microporous – there is no central chimney, but the rock itself is like a mineralised sponge, with thin walls separating interconnected pores, micrometres to millimetres in scale, altogether forming a vast labyrinth through which the alkaline hydrothermal fluids percolate (**Figure 13**). Because these fluids are not superheated by magma, their temperatures favour not only the synthesis of organic molecules (more on this soon) but also slower rates of flow. Rather than being pumped out at a furious speed, the fluids wend their way gently across catalytic surfaces. And the vents persist for millennia, at least 100,000 years in the case of Lost City. As Mike Russell points out, that's 10^{17} microseconds, a more meaningful time unit for measuring chemistry. Time aplenty.

Thermal currents through microporous labyrinths have a remarkable capacity to concentrate organic molecules (including amino acids, fatty acids and nucleotides) to extreme levels, thousands or even millions of times the starting concentration, by way of a process known as thermophoresis. This is a little like the tendency of small items of laundry to accumulate inside a duvet cover in the washing machine. It all depends on kinetic energy. At higher temperatures, small molecules (and small items of laundry) dance around, with some freedom to move in all directions. As the hydrothermal fluids mix and cool, the kinetic energy of the organic molecules falls, and their freedom to dance around diminishes (which is what happens to socks inside the duvet cover). That means they are less likely to leave again, and so they accumulate in these regions of lower kinetic energy (**Figure 13**). The power of thermophoresis depends in part on molecular size: large molecules, such as nucleotides, are retained better than smaller ones. Small end products, such as methane, are easily lost from the vent. All in all, continuous hydrothermal flow through microporous vents should actively concentrate organics by a dynamic process that does not alter the steady-state conditions (unlike freezing or evaporation) but actually *is* the steady state. Better still, thermophoresis drives the formation of dissipative structures within vent pores, by promoting interactions between organics. These can spontaneously precipitate fatty acids into vesicles, and possibly polymerise amino acids and nucleotides into proteins and RNA. Such interactions are a matter of concentration: any process that increases concentration promotes chemical interactions between molecules.

This may sound too good to be true, and in one sense it is. The alkaline hydrothermal vents at Lost City are home to plenty of life today, albeit mostly rather undramatic bacteria and archaea. They also produce low

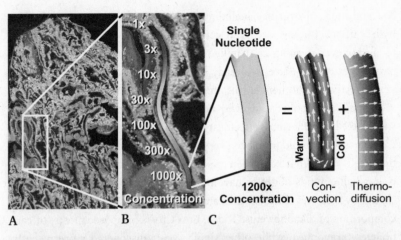

A B C

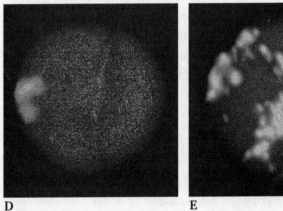

D E

Figure 13 **Extreme concentration of organics by thermophoresis**
A A section of an alkaline hydrothermal vent from Lost City, showing the porous structure of the walls – there is no central chimney but an interconnected labyrinth of pores, micrometres to millimetres in diameter. **B** Organics such as nucleotides can theoretically concentrate up to more than 1000 times their starting concentration by thermophoresis, driven by convection currents and thermal diffusion in the vent pores, illustrated in **C**. **D** An example of experimental thermophoresis from our reactor at University College London, showing 5000-fold concentration of a fluorescent organic dye (fluorescein) in a microporous ceramic foam (diameter: 9 cm). **E** Even greater concentration of the fluorescent molecule quinine, at least 1 million-fold in this case.

concentrations of organics, including methane and trace amounts of other hydrocarbons. But these vents are certainly not giving rise to new life forms today, nor even forming a rich milieu of organics by thermophoresis. That's partly because the bacteria already living there hoover up any resources very effectively; but there are also more fundamental reasons.

Just as black smoker vents were not exactly the same 4 billion years ago, so alkaline hydrothermal vents must have been different in their chemistry. Certain aspects would have been very similar. The process of serpentinisation itself should not have been any different: the same warm, hydrogen-rich, alkaline fluids ought to have bubbled up to the sea floor. But ocean chemistry was very different then and that should have altered the mineral composition of alkaline vents. Today, Lost City is composed mostly of carbonates (aragonite), while other similar vents discovered more recently (such as Strýtan, in northern Iceland) are composed of clays. Back in the Hadean oceans, 4 billion years ago, we can't be sure what kind of structures would have formed, but there were two main differences that must have had a big effect: oxygen was absent, and CO_2 concentration in the air and ocean was much greater. These differences should have made ancient alkaline vents far more effective as flow reactors.

In the absence of oxygen, iron dissolves in its ferrous form. We know that the early oceans were full of dissolved iron, because later on it all precipitated out as vast banded iron formations, as noted in Chapter 1. Much of this dissolved iron came from black smoker (volcanic) vents. We also know that iron would have precipitated out in alkaline hydrothermal vents – not because we have seen it, but because the rules of chemistry dictate it; and we can simulate it in the lab. In this case, the iron would have precipitated as iron hydroxides and iron sulphides, which form catalytic clusters that are still found in enzymes driving carbon and energy metabolism today – proteins like ferredoxin. In the absence of oxygen, then, the mineral walls of alkaline vents would have contained catalytic iron minerals, likely doped with other reactive metals such as nickel and molybdenum (which dissolves in alkaline fluids). Now we are getting close to a real flow reactor: hydrogen-rich fluids circulate through a labyrinth of micropores with catalytic walls that concentrate and retain products while venting waste.

But what exactly is reacting? Here we are reaching the crux of the matter. This is where the high CO_2 levels enter the equation. The alkaline hydrothermal vents of today are relatively starved of carbon, because much of the available inorganic carbon precipitates out as carbonate (aragonite) in the vent walls. Back in Hadean times, 4 billion years ago, our best guess is that CO_2 levels were substantially higher, perhaps 100–1,000 times greater than today. Beyond relieving the carbon limitation of primordial vents, high CO_2 levels would also have made the oceans more acidic, in turn making it harder to precipitate calcium carbonate. (This is threatening coral reefs today, as rising CO_2 begins to acidify modern oceans.) The pH of modern oceans is around 8, slightly alkaline. In the Hadean, the oceans are likely to have been neutral or mildly acidic, perhaps pH 5–7, although the actual value is practically unconstrained by geochemical proxies. The combination of high CO_2, mildly acidic oceans, alkaline fluids, and thin, FeS-bearing vent walls is crucial, because it promotes chemistry that would otherwise not happen easily.

Two broad principles govern chemistry: thermodynamics and kinetics. Thermodynamics determines which states of matter are more stable – which molecules will form, given unlimited time. Kinetics relates to speed – which products will form in a limited time. In terms of thermodynamics, CO_2 will react with hydrogen (H_2) to form methane (CH_4). This is an exothermic reaction, meaning that it releases heat. That in turn increases the entropy of the surroundings, at least under certain conditions, favouring the reaction. Given the opportunity, it should occur spontaneously. The conditions required include moderate temperatures and an absence of oxygen. If the temperature climbs too high, CO_2 is more stable than methane, as already noted. Likewise, if oxygen is present, it will react preferentially with hydrogen to form water. Four billion years ago, the moderate temperatures and anoxic conditions in alkaline vents should have favoured the reaction of CO_2 with H_2 to form CH_4. Even today, with some oxygen present, Lost City produces a small amount of methane. The geochemists Jan Amend and Tom McCollom have gone even further and calculated that the formation of organic matter from H_2 and CO_2 is thermodynamically favoured under alkaline hydrothermal conditions, so long as oxygen is excluded. That's remarkable. Under these conditions, between 25 and 125°C, the formation of total cell biomass (amino acids, fatty acids, carbohydrates, nucleotides and so on) from H_2 and CO_2 is actually *exergonic*. This means that organic matter should form spontaneously from H_2 and CO_2 under these conditions. The formation of cells releases energy and increases overall entropy!

But – and this is a big but – H_2 does not easily react with CO_2. There is a *kinetic* barrier, meaning that although thermodynamics says they should react spontaneously, some other obstacle stops it from happening right away. H_2 and CO_2 are practically indifferent to each other. To force them to react together requires an input of energy – a firecracker to break the ice. Now they will react, initially to form partially reduced compounds. CO_2 can only accept electrons in pairs. Addition of two electrons gives formate ($HCOO^-$); two more give formaldehyde (CH_2O); another two give methanol (CH_3OH); and a final pair gives the fully reduced methane (CH_4). Life, of course, is not made of methane, but it is only partially reduced carbon, roughly equivalent in its redox state to a mixture of formaldehyde and methanol. This means there are two important kinetic barriers relating to the origins of life from CO_2 and H_2. The first needs to be overcome, to get to formaldehyde or methanol. The second must *not* be overcome! Having coaxed H_2 and CO_2 into a warm embrace, the last thing a cell needs is for the reaction to run straight through to methane. Everything would dissipate and disperse as a gas, and that would be that. Life, it seems, knows exactly how to lower the first barrier and exactly how to keep the second barrier raised (only dropping it when it needs the energy). But what happened at the beginning?

This is the stumbling point. If it were easy to get CO_2 to react with H_2 economically – without putting in more energy than we get out – then we would have done it by now. That would be a huge step to solving the world's energy problems. Imagine it: mimic photosynthesis to split water, releasing H_2 and O_2. That's been done, and could potentially drive a hydrogen economy. But there are practical drawbacks to a hydrogen economy. How much better to react the H_2 with CO_2 from the air to make natural gas or even synthetic gasoline! Then we can go right on burning gas in our power stations. That would balance CO_2 emissions with CO_2 capture, halting the rise in atmospheric CO_2 levels and relieving our dependence on fossil fuels. Energy security. The returns could hardly be greater, and yet we have still not succeeded in driving this simple reaction economically. Well ... that's what the simplest living cells do all the time. Methanogens, for example, get all the energy and all the carbon needed to grow from reacting H_2 with CO_2. But more difficult: how could it have been done *before* there were any living cells? Wächtershäuser dismissed this as impossible: life could not have started from the reaction of CO_2 and H_2, he said, they simply won't react.[8] Even ramping up the pressure to the intense pressures found several kilometres down in hydrothermal vents at the bottom of the oceans does not force H_2 to react with CO_2. That's why Wächtershäuser came up with the idea of 'pyrites pulling' in the first place.

But there is one possible way.

Proton power

Redox reactions involve the transfer of electrons from a donor (H_2 in this case) to an acceptor (CO_2). The willingness of a molecule to transfer its electrons is connoted in the term 'reduction potential'. The convention is not helpful, but is easy enough to understand. If a molecule 'wants' to be rid

8 I'm sad to say that this is now the considered view of Mike Russell too. He has tried and failed to force CO_2 to react with H_2 to produce formaldehyde and methanol, and no longer believes it is possible. In collaboration with Wolfgang Nitschke, he now calls on other molecules, notably methane (produced in vents) and nitric oxide (arguably present in early oceans) to drive the origin of life, via a process analogous to modern methanotrophic bacteria. Bill Martin and I disagree with them, for reasons that I won't discuss here but, if you are really interested, you will find in Sousa *et al.*, given in Further Reading. This is not a trivial question, as it depends on the oxidation state of the early oceans, but it is amenable to experimental testing. A major advance over the past decade or so is precisely that the alkaline vent theory is now being considered very seriously by a widening group of scientists, who are formulating specific and distinct testable hypotheses within a similar overall framework, and setting out to test them experimentally. This is how science should work, and I don't doubt that all of us would be happy to be proved wrong on details, while hoping (naturally) that the overall framework stands robust.

of its electrons, it is assigned a negative value; the more that it wants to be rid of its electrons, the more negative is the reduction potential. Conversely, if an atom or molecule craves electrons and will pick them up from almost anywhere, it is assigned a positive value (you could think of it as the power of attraction for negatively charged electrons). Oxygen 'wants' to grab electrons (oxidising whatever it takes them from), giving it a very positive reduction potential. All these terms are in fact relative to the so-called standard hydrogen electrode, but we don't need to worry about that here.[9] The point is that a molecule with a negative reduction potential will tend to get rid of its electrons, passing them on to any molecule with a more positive reduction potential, but not the other way around.

That's the problem with H_2 and CO_2. At neutral pH (7.0), the reduction potential of H_2 is technically −414 mV. If H_2 gives up its two electrons, that leaves behind two protons, $2H^+$. The reduction potential for hydrogen reflects this dynamic balance – the tendency of H_2 to lose its electrons, becoming H^+, and the tendency of $2H^+$ to pick up electrons to form H_2. If CO_2 were to pick up those electrons, it would become formate. But formate has a reduction potential of −430 mV. That means it will tend to pass electrons on to H^+ to form CO_2 and H_2. Formaldehyde is even worse. Its reduction potential is about −580 mV. It is extremely reluctant to hang on to its

---

9 OK, so you are worried… Reduction potential is measured in millivolts. Imagine an electrode made of magnesium inserted into a beaker of magnesium sulphate solution. Magnesium has a strong tendency to ionise, releasing more Mg^{2+} ions into the solution, and leaving electrons behind on the electrode. That imparts a negative charge, which can be quantified relative to a standard 'hydrogen electrode'. This is an inert platinum electrode, in an atmosphere of hydrogen, which is inserted into a solution of protons at pH 0 (1 gram of protons per litre) and 25°C. If the magnesium and standard hydrogen electrodes are connected by a wire, electrons will flow from the negative magnesium electrode to the relatively positive (in fact it's just less negative) hydrogen electrode, to form hydrogen gas, by abstracting protons from the acid. Magnesium actually has a very negative reduction potential (−2.37 volts, to be precise) compared with the standard hydrogen electrode. Notice that all these values are at pH 0, by the way. In the main text I say that the reduction potential of hydrogen is −414 mV at pH 7. That's because the reduction potential gets more negative by about −59 mV for each pH unit increase (see main text).

electrons, and will easily pass them on to protons to form H_2. Thus when considering pH 7, Wächtershäuser is correct: there is no way that H_2 can reduce CO_2. But of course some bacteria and archaea live from exactly this reaction, so it must be possible. We'll look into the details of how they do that in the next chapter, as they are more relevant to the next stage of our story. For now, all we need to know is that bacteria growing from H_2 and CO_2 can *only* grow when powered by a proton gradient across a membrane. And that's a helluva clue.

The reduction potential of a molecule often depends on pH, which is to say on proton concentration. The reason is simple enough. Transfer of an electron transfers a negative charge. If the molecule that is reduced can also accept a proton, the product becomes more stable, as the positive charge of the proton balances the negative charge of the electron. The more protons available to balance charges, the more easily an electron transfer will proceed. That makes the reduction potential more positive – it becomes easier to accept a pair of electrons. In fact, the reduction potential increases by about 59 mV for each pH unit of acidity. The more acidic the solution, the easier it is to transfer electrons on to CO_2 to produce formate or formaldehyde. Unfortunately, exactly the same applies to hydrogen. The more acidic the solution, the easier it is to transfer electrons on to protons to form H_2 gas. Simply changing pH therefore has no effect at all. It remains impossible to reduce CO_2 with H_2.

But now think of a proton gradient across a membrane. The proton concentration – the acidity – is different on opposite sides of the membrane. Exactly the same difference is found in alkaline vents. Alkaline hydrothermal fluids wend their way through the labyrinth of micropores. So do mildly acidic ocean waters. In some places there is a juxtaposition of fluids, with acidic ocean waters saturated in CO_2 separated from alkaline fluids rich in H_2 by a thin inorganic wall, containing semiconducting FeS minerals. The reduction potential of H_2 is lower in alkaline conditions: it desperately 'wants' to be rid of its electrons, so the left-over H^+ can pair up with the OH^- in the alkaline fluids to form water, oh so stable. At pH 10, the reduction potential of H_2 is −584 mV: strongly reducing. Conversely, at pH 6, the reduction potential for formate is −370m V, and for formaldehyde it is

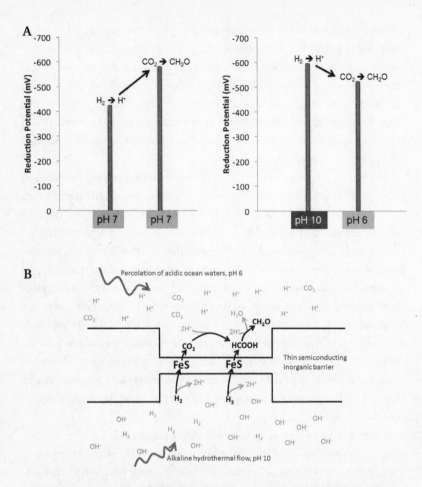

Figure 14 **How to make organics from H₂ and CO₂**

A The effect of pH on reduction potential. The more negative the reduction potential, the more likely a compound is to transfer one or more electrons; the more positive, the more likely it is to accept electrons. Note that the scale on the Y axis becomes more negative with height. At pH 7, H_2 can't transfer electrons to CO_2 to produce formaldehyde (CH_2O); the reaction would rather proceed in the opposite direction. However, if H_2 is at pH 10, as in alkaline hydrothermal vents, and CO_2 is at pH 6, as in early oceans, the reduction of CO_2 to CH_2O is theoretically possible. **B** In a microporous vent, fluids at pH 10 and pH 6 could be juxtaposed across a thin semiconducting barrier containing FeS minerals, facilitating the reduction of CO_2 to CH_2O. FeS is here acting as a catalyst, as it still does in our own respiration, transferring electrons from H_2 to CO_2.

−520m V. In other words, given this difference in pH, it is quite easy for H_2 to reduce CO_2 to make formaldehyde. The only question is: how are electrons physically transferred from H_2 to CO_2? The answer is in the structure. FeS minerals in the thin inorganic dividing walls of microporous vents conduct electrons. They don't do it nearly as well as a copper wire, but they do it, nonetheless. And so in theory, the physical structure of alkaline vents should drive the reduction of CO_2 by H_2, to form organics (**Figure 14**). Fantastic!

But is it true? Here is the beauty of science. This is a simple testable question. That's not to say it is easy to test; I've been trying to do so in the lab, with the chemist Barry Herschy and PhD students Alexandra Whicher and Eloi Camprubi, for a while now. With funding from the Leverhulme Trust, we've built a small benchtop reactor to try to drive these reactions. Precipitating these thin, semiconducting FeS walls in the lab is not straightforward. There's also the problem that formaldehyde is not stable – it 'wants' to pass its electrons back on to protons, to form H_2 and CO_2 once again, and it will do that more easily in acidic conditions. The exact pH and hydrogen concentration is critical. And plainly it's not easy to simulate the colossal scale of real vents in the lab – tens of metres high, operating at intense pressures (which permits a much higher concentration of gases such as hydrogen). Yet despite all these issues, the experiment is simple in the sense that it is a circumscribed, testable question, an answer to which could tell us a great deal about the origin of life. And we have indeed produced formate, formaldehyde and other simple organics (including ribose and deoxyribose).

For now, let's take the theory at face value, and assume that the reaction will indeed take place as predicted. What will happen? There should be a slow but sustained synthesis of organic molecules. We'll discuss which ones, and how exactly they should be formed, in the next chapter; for now let's just note that this is another simple testable prediction. Once formed, these organics should be concentrated to thousands of times their starting concentration by thermophoresis, as discussed earlier, promoting the formation of vesicles and perhaps polymers like proteins. Once again, the predictions that organics will concentrate and then polymerise are testable directly in the lab, and we are trying to do so. First steps are encouraging:

the fluorescent dye fluorescein, similar in size to a nucleotide, concentrates at least 5000-fold in our through-flow reactor, and quinine may concentrate even more (**Figure 13**).

So what does all this stuff about reduction potentials really mean? It at once constrains and opens wide the conditions under which life should evolve in the universe. This is one of the reasons that scientists often look as if they are in their own little world, lost in abstract thought about the most arcane details. Can there possibly be any mighty import about the fact that the reduction potential of hydrogen falls with pH? Yes! Yes! Yes! Under alkaline hydrothermal conditions, H_2 should react with CO_2 to form organic molecules. Under almost any other conditions, it will not. In this chapter, I have already ruled out virtually all other environments as workable settings for the origin of life. We have established on thermodynamic grounds that to make a cell from scratch requires a continuous flow of reactive carbon and chemical energy across rudimentary catalysts in a constrained through-flow system. Only hydrothermal vents provide the requisite conditions, and only a subset of vents – alkaline hydrothermal vents – match all the conditions needed. But alkaline vents come with both a serious problem and a beautiful answer to the problem. The serious problem is that these vents are rich in hydrogen gas, but hydrogen will not react with CO_2 to form organics. The beautiful answer is that the physical structure of alkaline vents – natural proton gradients across thin semiconducting walls – will (theoretically) drive the formation of organics. And then concentrate them. To my mind, at least, all this makes a great deal of sense. Add to this the fact that all life on earth uses (still uses!) proton gradients across membranes to drive both carbon and energy metabolism, and I'm tempted to cry, with the physicist John Archibald Wheeler, 'Oh, how could it have been otherwise! How could we all have been so blind for so long!'

Let's calm down and finish. I said that reduction potentials both constrain and open the conditions under which life should evolve. By this analysis, the conditions that best encourage the origins of life are found in alkaline vents. Perhaps your heart is sinking ... why narrow down the options so tightly? Surely there must be other ways! Well, maybe. In an infinite universe, anything is possible; but that doesn't make it probable. Alkaline vents are probable. They are formed, remember, by a chemical reaction between water and the mineral olivine. Rock. In fact, one of the most abundant minerals in the universe, a major part of interstellar dust and the accretion discs from which planets, including the earth, are formed. Serpentinisation of olivine may even occur in space, hydrating the interstellar dust. When our planet accreted, this water was driven off by the rising temperatures and pressures, giving rise, some say, to the earth's oceans. However that may be, olivine and water are two of the most abundant substances in the universe. Another is CO_2. This is a common gas in the atmosphere of most planets in the solar system, and has even been detected in the atmosphere of exoplanets in other stellar systems.

Rock, water and CO_2: the shopping list for life. We will find them on practically all wet rocky planets. By the rules of chemistry and geology, they will form warm alkaline hydrothermal vents, with proton gradients across thin-walled catalytic micropores. We can count on that. Perhaps their chemistry is not always conducive to life. Yet this is an experiment going on right now, on as many as 40 billion earth-like planets in the Milky Way alone. We live in a cosmic culture dish. How often these perfect conditions give rise to life depends on what happens next.

4

THE EMERGENCE OF CELLS

$\mathcal{O}$

'I think', wrote Darwin: just those two words, scrawled next to a sketch
of a branching tree of life, in a notebook from 1837. That was only a
year after returning from the voyage of the *Beagle*. Twenty-two years later,
a more artfully drawn tree was the only illustration in *The Origin of Species*.
The idea of a tree of life was so central to Darwin's thinking, and to the
currency of evolutionary biology ever since, that it's shocking to be told it
is wrong, as *New Scientist* did in large letters on their front cover in 2009,
150 years after the publication of Darwin's *Origin*. The cover flirted shamelessly with an extended readership, but the article itself was more moderate
in tone and made a specific point. To a degree that is very hard to define, the
tree of life is indeed wrong. That does not mean Darwin's major contribu-
tion to science, evolution by natural selection, is also wrong: it merely
shows that his knowledge of heredity was limited. That's not news. It is
well known that Darwin knew nothing of DNA, or genes, or Mendel's laws,
let alone the transfer of genes between bacteria, so his view of heredity was
through a glass, darkly. None of that discredits Darwin's theory of natural
selection; hence the cover was correct in a narrow technical sense, but
grossly misleading in a deeper sense.

What the cover did do, though, was bring a serious issue to the fore-
ground. The idea of a tree of life assumes 'vertical' inheritance, in which
parents pass on copies of genes to their offspring by sexual reproduction.

Over generations, genes are passed on mostly within a species, with relatively little intercourse between species. Populations that become reproductively isolated diverge slowly over time, as the interactions between them decline, and ultimately form new species. This gives rise to the branching tree of life. Bacteria are more equivocal. They don't have sex in the eukaryotic way, so they don't form nice neat species in the same way either. Defining the term 'species' in bacteria has always been problematic. But the real difficulty with bacteria is that they spread their genes around by 'lateral' gene transfer, passing handfuls of genes from one to another like small change, as well as bequeathing a copy of their full genome to daughter cells. None of this undermines natural selection in any sense – it is still descent with modification; it's just that the 'modification' is achieved in more ways than we once thought.

The prevalence of lateral gene transfer in bacteria poses a deep question about what we *can* know – a question as fundamental in its own way as the celebrated 'uncertainty principle' in physics. Almost any tree of life you look at from the modern era of molecular genetics will be based on a single gene, chosen carefully by the pioneer of molecular phylogenetics, Carl Woese – the gene for small subunit ribosomal RNA.[1] Woese argued (with some justification) that this gene is universal across life, and is rarely, if ever, transferred by lateral gene transfer. It therefore supposedly indicates the 'one true phylogeny' of cells (**Figure 15**). In the limited sense that one cell gives rise to daughter cells, and that these daughter cells are always likely to share the ribosomal RNA of their parent, this is true. But what happens if, over many generations, other genes are replaced by lateral gene

---

1 See the Introduction. The ribosomes are the protein-building factories found in all cells. These large molecular complexes have two major subunits (large and small), which are themselves composed of a mixture of proteins and RNA. The 'small subunit ribosomal RNA' is what Woese sequenced, in part because it was fairly easy to extract (there are thousands of ribosomes in any one cell); and in part because protein synthesis is fundamental to life, and so is universally conserved with only trivial differences between humans and hydrothermal bacteria. It is never easy to replace the foundation stones of any building or discipline; and for much the same reasons, ribosomes are rarely transferred between cells.

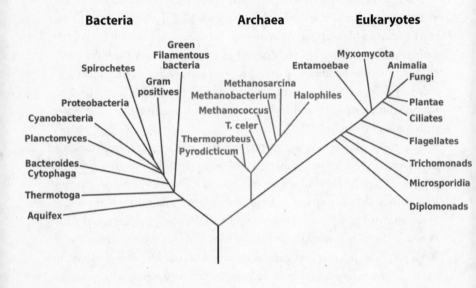

Bacteria   **Archaea**   **Eukaryotes**

Figure 15 **The famous but misleading three-domains tree of life**
The tree of life as portrayed by Carl Woese in 1990. The tree is based on a single highly conserved gene (for small subunit ribosomal RNA) and is rooted using the divergence between pairs of genes found in all cells (which must therefore have already been duplicated in the last universal common ancestor, LUCA). This rooting suggests that the archaea and the eukaryotes are more closely related to each other than either group is to the bacteria. However, while that is generally true for a core of informational genes, it is not true for the majority of genes in eukaryotes, which are more closely related to bacteria than archaea. This iconic tree is therefore profoundly misleading, and should be seen strictly as a tree of one gene only: it is most certainly not a tree of life!

transfer? In complex multicellular organisms, that rarely happens. We can sequence the ribosomal RNA of an eagle, and it will tell us that this is a bird. We can infer that it has a beak, feathers, claws, wings, lays eggs, and so on. That's because vertical inheritance ensures that there is always a good correlation between the ribosomal 'genotype' and the overall 'phenotype': the genes encoding all these birdlike traits are fellow travellers; they sail down the generations together, being modified over time, certainly, but rarely in a dramatic fashion.

But now imagine that lateral gene transfer predominates. So we sequence the ribosomal RNA, and it tells us we are dealing with a bird. Only now do we look at this 'bird'. It turns out to possess a trunk, six legs, eyes on its knees, fur; it produces eggs like frogspawn, lacks wings and howls like a hyena. Yes, of course that is absurd; but this is precisely the problem that we face with bacteria. Monstrous chimeras regularly stare us in the face; but because bacteria are typically small and morphologically simple, we don't scream. Nonetheless, in their genes bacteria are almost always chimeric, and some are real monsters, genetically as mangled as my 'bird'. Phylogeneticists really ought to scream. We can't infer what a cell might have looked like, or how it might have lived in the past, on the basis of its ribosomal genotype.

What is the use of sequencing a single gene if it can tell us nothing about the cell that it comes from? It can be useful, depending on the timescale and the rate of gene transfer. If the rate of lateral gene transfer is low (as in plants and animals, many protists, and some bacteria) then there will be a good correlation between ribosomal genotype and phenotype, so long as we are careful not to range too far back into the past. But if the rate of gene transfer is fast, that correlation can be wiped out very quickly. The difference between pathogenic variants of *E. coli* and harmless common strains is not reflected in the ribosomal RNA, but in the acquisition of other genes that confer aggressive growth – as much as 30% of the genome can vary in different strains of *E. coli* – that's 10 times the variation between us and chimpanzees, yet we still call them the same species! A ribosomal RNA phylogeny is the last thing you need to know about these killer bugs. Conversely, prolonged periods of time will wipe out a correlation even if the rates of lateral gene transfer are very low. That means it is almost impossible to know how a bacterium earned its living 3 billion years ago, given that slow rates of transfer could have replaced essentially all of its genes many times over in that period.

And so the conceit behind the tree of life is wrong. The hope is that it may be possible to reconstruct the one true phylogeny of all cells, to infer how one species arose from another, to trace relatedness back to the beginning, ultimately allowing us to infer the genetic make-up of the common ancestor of all life on earth. If we could indeed do that, we would know everything about that last ancestral cell, from its membrane composition, to the environment in which it lived, to the molecules that fuelled its growth. But we cannot know these things with that precision. A striking test was laid out by Bill Martin, in a visual paradox which he calls the 'amazing disappearing tree'. He considered 48 genes that are universally conserved across all life, and built a gene tree for each of these genes, to show the relationship between 50 bacteria and 50 archaea (**Figure 16**).[2] At the tips of this tree, all 48 genes recovered exactly the same relationship between all 100 species of bacteria and archaea. Likewise at the very base: nearly all 48 genes 'agreed' that the deepest branch in the tree of life is between the bacteria and the archaea. In other words, the last universal common ancestor, fondly known as LUCA, was the common ancestor of bacteria and archaea. But when it comes to elucidating the deep branches within either bacteria or archaea, not a single gene tree could agree. All 48 genes gave a different tree! The problem could be technical (the signal is eroded by sheer distance) or the result of lateral gene transfer – patterns of vertical descent are destroyed if individual genes are swapped around at random. We don't know which possibility is true, and at the moment it seems impossible to say.

What does that mean? In essence, it means we can't determine which species of bacteria or archaea are the most ancient. If one gene tree says that methanogens are the most ancient archaea, the next tree says they are not,

---

2 Recall that the bacteria and archaea are the two great domains of prokaryotes, which are very similar in their morphological appearance but differ fundamentally in aspects of their biochemistry and genetics.

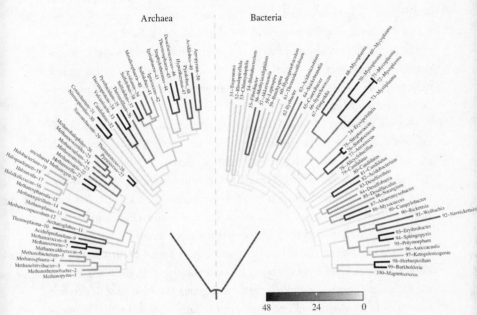

Figure 16 **The 'amazing disappearing tree'**
The tree compares the branching of 48 universally conserved genes in 50 bacteria and 50 archaea. All 48 genes are concatenated into a single sequence, giving greater statistical power (a common practice in phylogenetics); this 'supergene' sequence is next used to build a tree showing how the 100 species relate to each other. Each individual gene is then used to build a separate tree, and each of these trees is compared with the 'supergene' tree built from the concatenated genes. The strength of shading denotes the number of individual gene trees that correspond to the concatenated tree for each branch. At the base of the tree, nearly all the 48 genes recover the same tree as the concatenated sequence, clearly indicating that the archaea and the bacteria are genuinely deeply divided. At the tips of the branches, most individual gene trees also agree with the concatenated tree. But the deeper branches within both groups have vanished: not a single individual gene tree recovers the same branching order as the concatenated sequence. This problem could be a result of lateral gene transfers confounding branching patterns, or simply the erosion of a statistically robust signal over an unimaginable 4 billion years of evolution.

so it is practically impossible to reconstruct which properties the most ancient cells might have had. Even if by some ingenious means we could prove that methanogens are indeed the most ancient archaea, we still could not be sure that they always lived by making methane, as do modern methanogens. Lumping genes together to enhance the signal strength does not help much as each gene may have had a different history, making any composite signal a fabrication.

But the fact that all 48 of Bill Martin's universal genes agree that the deepest divergence in the tree of life is between bacteria and archaea holds out some hope. If we can figure out which properties are shared by all bacteria and archaea, and which are distinct, presumably arising later in particular groups, then we can put together a 'photo fit' of LUCA. Yet here again we quickly run into trouble: genes found in both archaea and bacteria could have arisen in one group and been transferred into the other group by lateral gene transfer. Transfers of genes across entire domains are well known. If such transfers occurred early in evolution – in the blank bits of the amazing disappearing tree – then these genes would appear to descend vertically from a common ancestor, even though they did not. The more useful the gene, the more likely it is to have been transferred widely, early on in evolution. To discount such widespread lateral gene transfer, we are obliged to fall back on genuinely universal genes, which are shared by representatives of essentially every group of bacteria and archaea. That at least minimises the possibility that these genes were passed around by early lateral gene transfer. The problem now is that there are fewer than 100 such universal genes, a remarkably small number, and they paint a very peculiar picture of LUCA.

We've already noted this strange portrait in Chapter 2. Taken at face value, LUCA had proteins and DNA: the universal genetic code was already in operation, DNA was read off into transcripts of RNA, and then translated into proteins on ribosomes, those mighty molecular factories that build proteins in all known cells. The remarkable molecular machinery needed for reading off DNA, and for protein synthesis, is composed of scores of proteins and RNAs common to both bacteria and archaea. From their structures and sequences these machines appear to have diverged very early in evolution, and have not been swapped around much by lateral gene transfer. So far so good. Equally, bacteria and archaea are all chemiosmotic, driving ATP synthesis using proton gradients across membranes. The ATP synthase enzyme is another extraordinary molecular machine, on a par with the ribosome itself and apparently sharing its antiquity. Like the ribosome, the ATP synthase is universally conserved across all life, but differs in a few details of its structure in bacteria and archaea, suggesting that it diverged from a common ancestor in LUCA, without much confounding lateral gene transfer later on. So the ATP synthase, like ribosomes, DNA and RNA, seems to have been present in LUCA. And then there are a few bits and pieces of core biochemistry, such as amino acid biosynthesis and parts of the Krebs cycle, which share common pathways in bacteria and archaea, again implying they were present in LUCA; but there is remarkably little else.

What's different? An astonishing parade. Most of the enzymes used for DNA replication are distinct in bacteria and archaea. What could be more fundamental than that! Possibly only the membrane – yet it, too, is distinct in bacteria and archaea. So is the cell wall. That means both of the barriers that separate living cells from their environment are utterly different in bacteria and archaea. It is almost impossible to guess exactly what their common ancestor might have possessed instead. The list goes on, but that will do. Of the six fundamental processes of living cells discussed in the previous chapter – carbon flux, energy flux, catalysis, DNA replication, compartmentalisation and excretion – only the first three share any deep similarity, and even then only in certain respects, as we shall see.

There are several possible explanations. LUCA could have possessed two copies of everything, and lost one copy in bacteria, and the other copy in archaea. That sounds inherently daft, but it can't be ruled out easily. For example, we know that mixtures of bacterial and archaeal lipids do make stable membranes; perhaps LUCA had both types of lipid, and her descendants later specialised by losing one or the other. That might conceivably be true for some traits, but is not generalisable to all, as it runs into a problem known as 'the genome of Eden'. If LUCA had everything, and her descendants streamlined later on, then she must have started out with an enormous genome, much larger than any modern prokaryote. That seems to me to put the cart before the horses – we have complexity before simplicity, and two solutions to every problem. And why did all the descendants lose one of everything? I don't buy it; roll on the second option.

The next possibility is that LUCA was a perfectly normal bacterium, with a bacterial membrane, cell wall and DNA replication. At some later point, one group of descendants, the first archaea, replaced all these traits as they adapted to extreme conditions such as high temperatures in hot vents. This is probably the most widely accepted explanation, but it too is hardly persuasive. If it is true, why are the processes of DNA transcription and translation into proteins so similar in bacteria and archaea, yet DNA replication so different? Why, if archaeal cell membranes and cell walls help archaea adapt to hydrothermal environments, did extremophile bacteria living in the same vents not replace their own membranes and walls with the archaeal versions, or something similar? Why do archaea living in the soil or open oceans not replace their membranes and walls with bacterial versions? Bacteria and archaea share the same environments across the world, yet remain fundamentally different in their genetics and biochemistry in all these environments, despite lateral gene transfer between the two domains. It's just not credible that all these profound differences could reflect adaptation to one extreme environment, and yet then remain fixed in archaea, without exception, regardless of how inappropriate they were for all other environments.

That leaves us with the final barefaced option. The apparent paradox is not a paradox at all: LUCA really was chemiosmotic, with an ATP synthase, but really did not have a modern membrane, or any of the large respiratory complexes that modern cells use to pump protons. She really did have DNA, and the universal genetic code, transcription, translation and ribosomes, but really had not evolved a modern method of DNA replication. This strange phantom cell makes no sense in an open ocean, but begins to add up when considered in the environment of alkaline hydrothermal vents discussed in the previous chapter. The clue lies in how bacteria and archaea live in these vents – some of them, at least, by an apparently primordial process called the acetyl CoA pathway, which bears an uncanny resemblance to the geochemistry of vents.

The rocky road to LUCA

Right across the entire living world, there are only six different ways of fixing carbon – of converting inorganic molecules such as carbon dioxide into organic molecules. Five of these pathways are quite complex and require an input of energy to drive them forwards, from the sun in photosynthesis for example. Photosynthesis is a good example for another reason too: the 'Calvin cycle', a biochemical pathway that traps carbon dioxide and converts it into organic molecules like sugars, is found only in photosynthetic bacteria (and plants, which acquired these bacteria as chloroplasts). This means that the Calvin cycle is unlikely to be ancestral. Had photosynthesis been present in LUCA, it must have been lost systematically from all archaea, plainly a foolish thing to do for such a useful trick. It is far more likely that the Calvin cycle arose later on, at the same time as photosynthesis, in the bacteria alone. Much the same goes for all but one of the other pathways too. Only one pathway of carbon fixation is found in both the bacteria and archaea, meaning that it plausibly arose in their common ancestor – the acetyl CoA pathway.

Even that claim is not quite true. There are some strange differences between bacteria and archaea in the acetyl CoA pathway, which we will address later in this chapter. For now, let's briefly consider the reasons why this pathway has a good claim to being ancestral, even though the phylogenetics are too ambiguous to support an early origin (neither do they discount it). The archaea that live by the acetyl CoA pathway are called methanogens, the bacteria acetogens. Some trees of life depict the methanogens as branching very deep; others depict the acetogens as branching very deep; and some depict both groups as evolving somewhat later on, with their simplicity purportedly reflecting specialisation and streamlining rather than an ancestral state. If we stick to phylogenetics alone, we may never be any the wiser. Luckily, we don't have to.

The acetyl CoA pathway starts with hydrogen and carbon dioxide – the same two molecules that we discussed in the last chapter as being plentiful in alkaline hydrothermal vents. As we noted then, the reaction between CO_2 and H_2 to form organics is exergonic, which is to say that it releases energy: in principle the reaction should take place spontaneously. In practice, there is an energetic barrier that prevents H_2 and CO_2 from reacting quickly. Methanogens use the proton gradient to overcome this barrier, which I shall argue was the ancestral state. Be that as it may, methanogens and acetogens both power their growth through the reaction of H_2 and CO_2 alone: that reaction provides all the carbon and all the energy needed for growth. This sets the acetyl CoA pathway apart from the other five pathways of carbon fixation. The geochemist Everett Shock summed it up memorably as 'a free lunch that you're paid to eat'. It may be a meagre lunch, but in the vents it's served all day.

That's not all. Unlike the other pathways, the acetyl CoA pathway is short and linear. There are fewer steps needed to get from simple inorganic molecules to the hub of metabolism in all cells, the small but reactive molecule acetyl CoA. Don't be afraid of words. CoA stands for coenzyme A, which is an important and universal chemical 'hook' to hang small molecules on, so that they can be processed by enzymes. The important bit is not so much the hook as what hangs from it, in this case the *acetyl* group. 'Acetyl' has the same root as acetic acid, vinegar, a simple two-carbon molecule that is at the centre of biochemistry in all cells. When attached to coenzyme A, the acetyl group is in an activated state (often called 'activated acetate' – in effect, reactive vinegar) which enables it to react readily with other organic molecules, thereby driving biosynthesis.

Thus the acetyl CoA pathway generates small reactive organic molecules from CO_2 and H_2, via just a few steps, while at once releasing enough energy to drive not just the formation of nucleotides and other molecules, but also their polymerisation into long chains – DNA, RNA, proteins, and so on. The enzymes that catalyse the first few steps contain inorganic clusters of iron, nickel and sulphur, which are physically responsible for transferring electrons on to CO_2 to form reactive acetyl groups. These inorganic clusters are basically minerals – rocks! – more or less identical in their structure to the iron–sulphur minerals that precipitate in hydrothermal vents (see **Figure 11**). The fit between the geochemistry of alkaline vents and the biochemistry of methanogens and acetogens is so close that the word analogous does not do it justice. Analogy implies similarity, which is potentially only superficial. In fact, the similarity here is so close that it might better be seen as true homology – one form physically gave rise to the other. So geochemistry gives rise to biochemistry in a seamless transition from the inorganic to the organic. As the chemist David Garner put it: 'It is the inorganic elements that bring organic chemistry to life.'[3]

But perhaps the greatest boon of acetyl CoA is that it sits at the crossroads of carbon and energy metabolism. The relevance of acetyl CoA to the origin of life was pointed out in the early 1990s by the distinguished Belgian biochemist Christian de Duve, albeit in the context of soup, not alkaline vents. Acetyl CoA not only drives organic syntheses, but it can also react directly with phosphate to form acetyl phosphate. While not as important an energy currency as ATP today, acetyl phosphate is still widely used across life, and can do much the same job as ATP. As noted in the last chapter, ATP does more than simply release energy; it also drives dehydration reactions, in which one molecule of water is extracted from two amino acids or other building blocks, thereby linking them together in a chain. The problem of dehydrating amino acids in solution, we noted, is equivalent to wringing out a wet cloth under water; but that is exactly what ATP does. We have shown in the lab that acetyl phosphate can do exactly the same job, as its chemistry is basically equivalent. This means that early carbon and energy metabolism could be driven by the same simple thioester, acetyl CoA.

Simple? I hear you say. The two-carbon acetyl group may be simple, but coenzyme A is a complex molecule, undoubtedly the product of natural selection, and therefore a later product of evolution. So is this whole argument circular? No, because there are genuinely simple 'abiotic' equivalents to acetyl CoA. The reactivity of acetyl CoA lies in its so-called 'thioester bond', which is no more than a sulphur atom bound to carbon, bound in turn to oxygen. It can be depicted as:

---

3 And the same inorganic elements still bring organic chemistry to life. More or less identical iron–sulphur clusters are found in our own mitochondria, more than a dozen of them in each respiratory chain (see **Figure 8** for complex I alone), meaning tens of thousands of them in every mitochondrion. Without them, respiration could not work and we would be dead in minutes.

$$R-S-CO-CH_3$$

in which 'R' stands for the 'rest' of the molecule, CoA in this case, and CH_3 is a methyl group. But the R needn't stand for CoA; it could stand for something as simple as another CH_3 group, giving a small molecule called methyl thioacetate:

$$CH_3-S-CO-CH_3$$

This is a reactive thioester, equivalent in its chemistry to acetyl CoA itself, but simple enough to be formed from H_2 and CO_2 in alkaline hydrothermal vents – indeed it has been produced by Claudia Huber and Günter Wächtershäuser from CO and CH_3SH alone. Even better, methyl thioacetate, like acetyl CoA, should be able to react directly with phosphate to form acetyl phosphate. And so this reactive thioester could in principle drive the synthesis of new organic molecules directly, as well as their polymerisation into more complex chains such as proteins and RNA, via acetyl phosphate – a hypothesis that we're testing in our benchtop reactor in the lab (in fact we have just succeeded in producing acetyl phosphate, albeit at low concentration).

A primordial version of the acetyl CoA pathway could in principle power everything needed for the evolution of primitive cells within the micropores of alkaline hydrothermal vents. I would envisage three stages. In the first stage, proton gradients across thin inorganic barriers containing catalytic iron–sulphur minerals drove the formation of small organic molecules (**Figure 14**). These organics were concentrated in the cooler vent pores by thermophoresis, and in turn acted as better catalysts, as we discussed in Chapter 3. These were the origins of biochemistry – the continuous formation and concentration of reactive precursors, fostering interactions between molecules and the formation of simple polymers.

The second stage was the formation of simple organic protocells within the pores of the vents, as a natural outcome of the physical interactions between organics – simple dissipative cell-like structures, formed by the self-organisation of matter, but as yet without any genetic basis or real complexity. I would see these simple protocells as depending on the proton gradient to drive organic synthesis, but now across their own organic membranes (lipid bilayers formed spontaneously from fatty acids, for example) rather than the inorganic walls of the vent itself. No proteins are needed for this. The proton gradient could drive the formation of methyl thioacetate and acetyl phosphate as discussed above, driving both carbon and energy metabolism. There's one key difference at this stage: new organic matter is now formed within the protocell itself, driven by natural proton gradients across organic membranes. Reading this back I'm struck by my overuse of the word 'drive'. It might be poor literary style, but there isn't a better word. I need to get across that this is not passive chemistry but it is *forced*, pushed, driven by the continuous flux of carbon, energy, protons. These reactions *need* to happen, they are the only way of dissipating the unstable disequilibrium of reduced, hydrogen-rich, alkaline fluids entering an oxidised, acidic, metal-rich ocean. The only way of reaching blessed thermodynamic equilibrium.

The third stage is the origin of the genetic code, true heredity, finally enabling protocells to make more or less exact copies of themselves. The earliest forms of selection, based on relative rates of synthesis and degradation, gave way to proper natural selection, in which populations of protocells with genes and proteins began to compete for survival within vent pores. The standard mechanisms of evolution eventually produced sophisticated proteins in early cells, including ribosomes and the ATP synthase, proteins conserved universally across life today. I envisage that LUCA, the common ancestor of bacteria and archaea, lived within the micropores of alkaline hydrothermal vents. That means all three stages from abiotic origins to LUCA take place within the vent pores. All are driven by proton gradients across inorganic walls or organic membranes; but the advent of sophisticated proteins such as the ATP synthase is a late step on this rocky road to LUCA.

I am not concerned in this book with the details of primordial biochemistry: where the genetic code came from, and other equally difficult problems. These are real problems, and there are ingenious researchers addressing them. We don't yet know the answers. But all these ideas assume a plentiful supply of reactive precursors. Just to give a single example, a beautiful idea from Shelley Copley, Eric Smith and Harold Morowitz on the origin of the genetic code posits that catalytic dinucleotides (two nucleotides joined together) could generate amino acids from simpler precursors, such as pyruvate. Their clever scheme shows how the genetic code may have arisen from deterministic chemistry. For those who are interested, I wrote a chapter on the origins of DNA in *Life Ascending*, which touched on some of these questions. But what all of these hypotheses take for granted is a steady supply of nucleotides, pyruvate, and other precursors. The question we are addressing here is: what were the driving forces that impelled the origin of life on earth? And my main point is simply that there is no *conceptual* difficulty about where all the carbon, energy and catalysts came from that drove the formation of complex biological molecules, right up to the advent of genes and proteins, and LUCA.

The vent scenario outlined here has a beautiful continuity with the biochemistry of methanogens, the archaea that live from H_2 and CO_2 by way of the acetyl CoA pathway. These apparently ancient cells generate a proton gradient across a membrane (we'll come to how they do that), reproducing exactly what alkaline hydrothermal vents provide for free. The proton gradient drives the acetyl CoA pathway, by way of an iron–sulphur protein embedded within the membrane – the energy-converting hydrogenase, or *Ech* for short. This protein funnels protons through the membrane on to another iron–sulphur protein, called ferredoxin, which in turn reduces CO_2. In the last chapter, I suggested that natural proton gradients across thin FeS walls in vents could reduce CO_2 by changing the reduction potentials of H_2 and CO_2. I suspect this is what *Ech* is doing on a nanometre scale. Enzymes often control the precise physical conditions (such as proton concentration) within clefts in the protein, just a few ångströms across, and *Ech* might be doing this too. If so, there could be an unbroken continuity between a primordial state, in which short polypeptides are stabilised by binding to FeS minerals embedded in fatty-acid protocells, and the modern state, in which the genetically encoded membrane protein *Ech* powers carbon metabolism in modern methanogens.

Be that as it may, the fact is that today, in the world of genes and proteins,

Ech draws on the proton gradient generated by methane synthesis to drive the reduction of CO_2. Methanogens also use the proton gradient to drive ATP synthesis directly, via the ATP synthase. Thus both carbon and energy metabolism are driven by proton gradients, exactly what the vents provided for free. The earliest protocells living in alkaline vents may have powered carbon and energy metabolism in precisely this way. That sounds plausible enough, but in fact relying on natural gradients brings its own problems. Intriguingly serious problems. Bill Martin and I realised there may be only one possible solution to these problems – and it gives a tantalising insight into why archaea and bacteria differ in fundamental ways.

The problem of membrane permeability

Inside our own mitochondria, the membranes are almost impermeable to protons. That's necessary. It is no good pumping protons across a membrane if they come rushing straight back at you, as if through innumerable little holes. You might as well try to pump water into a tank with a sieve for a base. In our mitochondria, then, we have an electrical circuit, in which the membrane works as an insulator: we pump protons across the membrane, and most of them return through proteins that behave as turbines, driving work. In the case of the ATP synthase, the flow of protons through this nanoscopic rotating motor drives ATP synthesis. But note that this whole system depends on active pumping. Block the pumps and everything grinds to a halt. That's what happens if we take a cyanide pill: it jams up the final proton pump of the respiratory chain in our mitochondria. If the respiratory pumps are impeded in this way, protons can continue to flow in through the ATP synthase for a few seconds before the proton concentration equilibrates across the membrane, and net flow ceases. It is almost as hard to define death as life, but the irrevocable collapse of membrane potential comes pretty close.

So how could a natural proton gradient drive ATP synthesis? It faces the 'cyanide' problem. Imagine a protocell sitting in a pore within a vent, powered by a natural proton gradient. One side of the cell is exposed to a continuous flow of ocean water, the other to a constant flux of alkaline hydrothermal flow (**Figure 17**). Four billion years ago, the oceans were probably mildly acidic (pH 5–7), while the hydrothermal fluids were equivalent to today, with a pH of about 9–11. Sharp pH gradients could therefore have been as much as 3–5 pH units in magnitude, which is to say that the difference in proton concentration could have been 1,000–100,000-fold.[4] For the sake of argument, imagine that the proton concentration inside the cell is similar to that of the vent fluids. That gives a difference in proton concentration between the inside and the outside, so protons will flow inwards down the concentration gradient. Within a few seconds, though, the influx should grind to a halt, unless the protons that flow in can be removed again. There are two reasons for this. First, the concentration difference swiftly evens out. And second, there is an issue with electrical charge. Protons (H^+) are positively charged, but in seawater their positive charge is counterbalanced by negatively charged atoms such as chloride ions (Cl^-). The problem is that protons cross the membrane much faster than chloride ions, so there is an influx of positive charge that is not offset by an influx of negative charge. The inside of the cell therefore becomes positively charged relative to the outside, and that opposes the influx of any more H^+. In short, unless there is a pump that can get rid of protons from inside the cell, natural proton gradients can't drive anything. They equilibrate, and equilibrium is death.

But there is an exception. If the membrane is nearly impermeable to protons, the influx must indeed cease. Protons enter the cell but can't leave again. But if the membrane is very leaky, it is a different story. Protons continue to enter the cell, as before, but now they can leave it again, albeit passively, through the leaky membrane on the other side of the cell. In effect, a leaky membrane imposes less of a barrier to flow. Better still,

---

4 Because the pH scale is logarithmic, 1 pH unit represents a 10-fold difference in proton concentration. Differences of this magnitude in such a small space may seem unfeasible, but in fact are possible because of the nature of fluid flow through pores on the scale of micrometres in diameter. Flow in these circumstances can be 'laminar', with little turbulence and mixing. The pore sizes in alkaline hydrothermal vents tend to combine both laminar and turbulent flow.

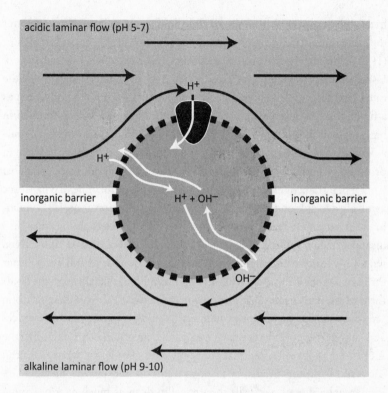

Figure 17 **A cell powered by a natural proton gradient**
A cell sits in the middle of the frame, enclosed by a membrane that is leaky to protons. The cell is 'wedged' in a small break in an inorganic barrier that separates two phases within a microporous vent. In the top phase, mildly acidic ocean water percolates along an elongated pore, at a pH of 5–7 (generally taken to be pH 7 in the model). In the bottom phase, alkaline hydrothermal fluids percolate along an unconnected pore, at a pH of about 10. Laminar flow indicates a lack of turbulence and mixing, which is characteristic of fluids flowing in small confined spaces. Protons (H^+) can flow directly through the lipid membrane, or through proteins embedded in the membrane (triangular shape), down a concentration gradient from the acidic ocean to the alkaline hydrothermal fluid. Hydroxide ions (OH^-) flow in the opposite direction, from the alkaline hydrothermal fluid to the acid ocean, but only through the membrane. The overall rate of proton flux depends on the permeability of the membrane to H^+, neutralisation by OH^- (to form H_2O); the number of membrane proteins; the size of the cell; and the charge across the membrane accrued by the movement of ions from one phase to the other.

hydroxide ions (OH^-) from the alkaline fluids cross the membrane at about the same rate as protons. When they meet, H^+ and OH^- react to form water (H_2O), eliminating the proton with its positive charge in one fell swoop. Using the classic equations of electrochemistry, it's possible to calculate the rates at which protons enter and exit a hypothetical (computational) cell as a function of membrane permeability. Victor Sojo, a chemist interested in big problems in biology, who is doing a PhD with me and Andrew Pomiankowski, has done exactly this. By tracking the steady-state difference in proton concentration, we could calculate the free energy (ΔG) available from a pH gradient alone. The results are just beautiful. The driving force available depends on the leakiness of the membrane to protons. If the membrane is extremely leaky, protons come rushing in like fools, but they also disappear again quickly, eliminated by a rapid influx of OH^- ions. Even with very leaky membranes, we found that protons will still enter faster through membrane proteins (like the ATP synthase) than through the lipids themselves. This means that proton flux can drive ATP synthesis or carbon reduction via the membrane protein *Ech*. Taking concentration differences and charge into consideration, as well as the operation of proteins like the ATP synthase, we showed that *only* cells with very leaky membranes can use natural proton gradients to power carbon and energy metabolism. Remarkably, these leaky cells theoretically glean as much energy from a natural proton gradient of 3 pH units as modern cells gain from respiration.

Actually, they could gain a lot more. Think again about methanogens. They spend most of their time generating methane, hence their name. On average, methanogens produce about 40 times as much waste (methane and water) as organic matter. All the energy derived from the synthesis of methane is used to pump protons (**Figure 18**). That's it. Methanogens spend practically 98% of their energy budget on generating proton gradients by methanogenesis, and little more than 2% producing new organic matter. With natural proton gradients and leaky membranes, none of that excessive energy spend is needed. The power available is exactly the same but the overheads are cut by at least 40-fold, a very substantial advantage. Just imagine having 40 times more energy! Even my young sons don't outdo me

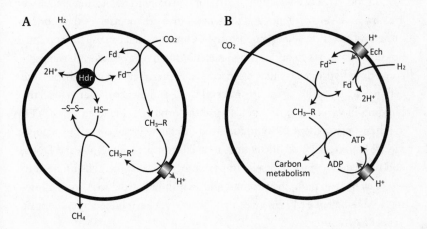

Figure 18 **Generating power by making methane**
A simplified view of methanogenesis. In **A** the energy from the reaction between H_2 and CO_2 powers the extrusion of protons (H^+) across the cell membrane. A hydrogenase enzyme (Hdr) catalyses the simultaneous reduction of ferredoxin (Fd) and a disulphide bond (–S–S–), using the two electrons from H_2. Ferredoxin in turn reduces CO_2, ultimately to a methyl (–CH$_3$) group bound to a cofactor designated R. The methyl group is then transferred to a second cofactor (R'), and this step releases enough energy to pump two H^+ (or Na^+) across the membrane. In the final stage, the –CH$_3$ group is reduced to methane (CH_4) by the HS– group. Overall, some of the energy released by the formation of methane (CH_4) from H_2 and CO_2 is conserved as an H^+ (or Na^+) gradient across the cell membrane. In **B** the H^+ gradient is used directly through two distinct membrane proteins to drive carbon and energy metabolism. The energy-converting hydrogenase (*Ech*) reduces ferredoxin (Fd) directly, which again passes its electrons on to CO_2 to form a methyl (–CH$_3$) group, which is reacted with CO to form acetyl CoA, the linchpin of metabolism. Likewise, H^+ flux through the ATP synthase drives ATP synthesis, and so energy metabolism.

that much. In the previous chapter, I mentioned that primitive cells would have needed *more* carbon and energy flux than modern cells; having zero need to pump gives them a lot more carbon and energy.

Consider a leaky cell, sitting in a natural proton gradient. Remember we are now in the era of genes and proteins, which are themselves the product of natural selection acting on protocells. Our leaky cell can use the continuous flux of protons to drive carbon metabolism, by way of *Ech*, the energy-converting hydrogenase that we discussed earlier. This protein enables the cell to react H_2 with CO_2, to form acetyl CoA, and thence onwards to all the building blocks of life. It can also use the proton gradient to drive ATP synthesis, using the ATP synthase. And of course it can use ATP to polymerise amino acids and nucleotides to make new proteins, RNA and DNA, and ultimately copies of itself. Importantly, our leaky cell has no need to waste energy on pumping protons, and so it should grow well, even allowing for inefficient early enzymes that had not yet been honed by billions of years of evolution.

But such leaky cells are also stuck right where they are, utterly dependent on hydrothermal flow and unable to survive anywhere else. When that flow ceases or shifts elsewhere, they are doomed. Even worse, they appear to be in an unevolvable state. There is no benefit to improving the properties of the membrane; on the contrary, less leaky membranes swiftly collapse the proton gradient, as there is no longer any way of getting rid of protons from inside the cell. So any variant cells that produced a more 'modern' impermeable membrane would be eliminated by selection. Unless they learnt to pump, of course; but that is equally problematic. We have seen that there is no point in pumping protons across a leaky membrane. Our study confirms that pumping offers no benefit, even if the permeability of the membrane is decreased by a whopping three orders of magnitude.

Let me spell that out. A leaky cell in a proton gradient has plenty of energy, enough to drive carbon and energy metabolism. If by some evolutionary sleight of hand, a fully functional pump is placed in the membrane, it offers no benefit whatsoever in terms of energy availability: the power available remains exactly the same as in its absence. That's because pumping protons over a leaky membrane is pointless – they come straight back through. Decrease membrane permeability by a factor of 10, and try again; still zero benefit. Decrease permeability by a factor of 100; still no benefit. Decrease permeability by a factor of 1,000; still no benefit. Why not? There is a balance of forces. Decreasing membrane permeability helps pumping, but also collapses the natural proton gradient, undermining the cell's power supply. Only if large amounts of pump are plastered across a nearly impermeable membrane (equivalent to that in our own cells) is there any benefit to pumping. That is a serious problem. There is no selective driving force for the evolution of either modern lipid membranes or modern proton pumps. Without a driving force they should not evolve; but they do exist, nonetheless. So what are we missing?

Here is an example of the serendipity of science. Bill Martin and I were pondering over exactly that question, and we mused that methanogens use a protein called an antiporter. The methanogens in question actually pump out sodium ions (Na^+), not protons (H^+), but they still have a few problems with protons accumulating inside. The antiporter swaps an Na^+ for an H^+, as if it were a strict two-way turnstile, or revolving door. For each Na^+ passing into the cell down a concentration gradient, one H^+ is forced out. It is a proton pump powered by a sodium gradient. But antiporters are pretty undiscriminating. They don't care which way round they work. If a cell pumped H^+ rather than Na^+, then the antiporter would simply go into reverse. For every H^+ that entered, one Na^+ would then be forced out. Ha! Suddenly we had it! If our leaky cell sitting in the alkaline hydrothermal vent evolved an Na^+/H^+ antiporter, it would act as a proton-powered Na^+ pump! For each H^+ that entered the cell through the antiporter, one Na^+ would be forced to leave! In theory, the antiporter could convert a natural proton gradient into a biochemical sodium gradient.

How would that help exactly? I should stress that this is a thought experiment, based on the known properties of the protein; but by our calculations, it could make a surprising difference. In general, lipid membranes are around six orders of magnitude less permeable to Na^+ than H^+. So a membrane that is extremely permeable to protons is fairly impermeable to sodium. Pump out a proton, and it will come straight back at you; pump out a sodium across the same membrane, and it won't come back nearly as fast.

This means that an antiporter can be driven by a natural proton gradient: for every H^+ that comes in, one Na^+ is extruded. So long as the membrane is leaky to protons, as before, proton flux through the antiporter will continue unabated, driving Na^+ extrusion. Because the membrane is less permeable to Na^+, the extruded Na^+ is more likely to stay out; or more specifically, it should re-enter the cell via membrane proteins, rather than coming straight back through the lipids. And that improves the coupling of Na^+ influx to work done.

Of course, that's only any use if the membrane proteins that drive carbon and energy metabolism – *Ech* and the ATP synthase – can't discriminate between Na^+ and H^+. That sounds preposterous, but rather surprisingly it may well be true. Some methanogens turn out to have ATP synthase enzymes that can be powered by either H^+ or Na^+, with roughly equal facility. Even the prosaic language of chemistry declares them to be 'promiscuous'. The reason could relate to the equivalent charge and very similar radii of the two ions. Although H^+ is much smaller than Na^+, protons rarely exist in isolation. When dissolved, they bind to water to form H_3O^+, which has a radius nearly identical to Na^+. Other membrane proteins including *Ech* are also promiscuous for H^+ and Na^+, presumably for the same reasons. The bottom line is that pumping Na^+ is by no means pointless. When powered by natural proton gradients, there is essentially no cost to extruding Na^+; and once a sodium gradient exists, Na^+ ions are more likely to re-enter the cell via membrane proteins such as *Ech* and ATP synthase than through membrane lipids. The membrane is now better 'coupled', meaning that it is better insulated, and therefore less likely to short-circuit. As a result, more ions are now available to drive carbon and energy metabolism, giving better payback for each ion pumped out.

There are several surprising ramifications of this simple invention. One is almost incidental: pumping sodium out of the cell lowers the concentration of sodium within the cell. We know that many core enzymes found in both bacteria and archaea (those responsible for transcription and translation, for example) have been optimised by selection to work at low Na^+ concentration, despite most probably evolving in the oceans, where the Na^+ concentration seems to have been high even 4 billion years ago. The early operation of an antiporter could potentially explain why all cells are optimised to low sodium, despite evolving in a high-sodium environment.[5]

More significantly for our immediate purposes, the antiporter effectively adds an Na^+ gradient to an existing H^+ gradient. The cell is still powered by the natural proton gradient, so still requires proton-permeable membranes; but it now has an Na^+ gradient too, which by our calculations gives the cell about 60% more power than it had before, when relying on protons alone. That gives cells two big advantages. First, cells with an antiporter have more power, and so can grow and replicate faster than cells without – an obvious selective advantage. Second, cells could survive on smaller proton gradients. In our study, cells with leaky membranes grow well with a proton gradient of about 3 pH units, which is to say that the proton concentration of the oceans (around pH 7) is three orders of magnitude greater than the proton concentration of alkaline fluids (about pH 10). By increasing the power of a natural proton gradient, cells with an antiporter could survive with a pH gradient of less than 2 pH units, allowing them to spread and colonise wider areas of the vent or contiguous vent systems. Cells with an antiporter would therefore tend to outcompete other cells, and would also spread and diverge in the vents. But because they still depend totally on the natural proton gradient, they could not leave the vents. One more step was required.

That brings us to the crucial point. With an antiporter, cells might not be able to leave the vent, but they are now primed to do so. In the parlance, an

---

5 The fact that ancient enzymes are optimised to a low Na^+/high K^+ content, given that the first membranes were leaky to these ions, can only mean that cells were optimised to the ionic balance of the surrounding medium, according to Russian bioenergeticist Armen Mulkidjanian. As the early oceans were high in Na^+, low in K^+, he believes that life can't have begun in the oceans. If he's right, then I must be wrong. Mulkidjanian points to terrestrial geothermal systems with high K^+, low Na^+, although these have problems of their own (he has organic synthesis driven by zinc sulphide photosynthesis, unknown in real life). But is it really impossible for natural selection to optimise proteins over 4 billion years, or are we to believe that the primordial ion balance was perfect for every enzyme? If it's possible to optimise enzyme function, how could that be done, given leaky early membranes? The use of antiporters in natural proton gradients offers a satisfying resolution.

antiporter is a 'preadaptation' – a necessary first step that facilitates a later evolutionary development. The reason is unexpected, or at least it was to me. For the first time, an antiporter favours the evolution of active pumping. I mentioned there is no benefit to pumping protons across a leaky membrane, because they come straight back at you. But with an antiporter there is an advantage. When protons are pumped out, some of them return not through the leaky lipids but through the antiporter, which extrudes Na^+ ions in their place. Because the membrane is better insulated to Na^+, more of the energy that had been spent on pumping out protons is retained as an ion gradient across the membrane. For every ion pumped out, there is a slightly higher chance it will stay out. And that means there is now a small advantage to pumping protons, whereas before there had been no advantage. Pumping only pays with an antiporter.

That's not all. Once a proton pump has evolved, there is now, for the first time, an advantage to improving the membrane. I reiterate: in a natural proton gradient it is strictly necessary to have a leaky membrane. Pumping protons across a leaky membrane is no use at all. An antiporter improves the situation because it increases the power available from a natural proton gradient, but it does not cut off the cell from its dependence on the natural gradient. Yet in the presence of an antiporter, it now pays to pump protons, meaning there is less dependence on the natural gradient. And now – only now! – is it better to have a less permeable membrane. Making the membrane slightly less leaky gives a slight advantage to pumping. Improving it a little bit more gives a slightly bigger advantage, and so on, all the way up to a modern proton-tight membrane. For the first time, we have a sustained selective driving force for the evolution of both proton pumps *and* modern lipid membranes. Ultimately, cells could cut their umbilical link to natural proton gradients: they were finally free to escape from the vents, and subsist in the great empty world.[6]

---

6 The alert reader may be wondering why cells don't just pump Na^+? It is indeed better to pump Na^+ across a leaky membrane than to pump H^+, but as the membrane becomes less permeable, that advantage is lost. The reason is esoteric. The power available to a cell depends on the concentration difference between the two sides of the membrane,

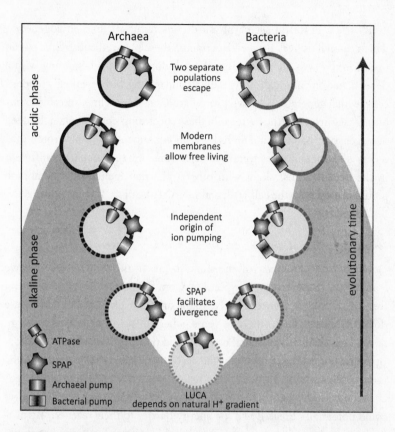

Figure 19 **The origin of bacteria and archaea**

A possible scenario for the divergence of bacteria and archaea, based on a mathematical model of energy availability in natural proton gradients. The figure shows only the ATP synthase for simplicity, but the same principle applies to other membrane proteins such as *Ech*. A natural H^+ gradient in a vent can drive ATP synthesis so long as the membrane is leaky (bottom), but there is no benefit to improving the membrane, as that collapses the natural gradient. A sodium–proton antiporter (SPAP) adds a biochemical sodium gradient to the geochemical proton gradient, enabling survival on smaller H^+ gradients, facilitating spread and divergence of populations in the vent. The extra power provided by SPAP means that pumping H^+ offers a benefit for the first time. With a pump, there is a benefit to lowering membrane permeability to H^+. When the membrane H^+ permeability approaches modern values, cells finally become independent of natural gradients, and can leave the vent. Bacteria and archaea are depicted escaping from the vent independently.

This is a beautiful set of physical constraints. Unlike phylogenetics, which can tell us very little with certainty, these physical constraints put an order on the possible succession of evolutionary steps, beginning with a dependence on natural proton gradients and ending with essentially modern cells, which generate their own proton gradients across impermeable membranes (**Figure 19**). And better still, these constraints could explain the deep divergence of bacteria and archaea. Both generate ATP using proton gradients across membranes, yet those membranes are fundamentally different in the two domains, along with other traits that include the membrane pumps themselves, the cell wall, and DNA replication. Let me explain.

Why bacteria and archaea are fundamentally different

Here is a brief summary of the story so far. In the previous chapter we considered, from an energetic point of view, the possible environments on the early earth that were conducive to the origin of life. We narrowed these down to alkaline hydrothermal vents, where a steady carbon and energy flux is combined with mineral catalysts and natural compartmentalisation. But these vents face a problem: the carbon and energy flux comes in the form of H_2 and CO_2, which do not react together easily. We saw that geochemical proton gradients across thin semiconducting barriers in vent pores could potentially break down the energy barrier to their reaction. By producing reactive thioesters such as methyl thioacetate (functionally equivalent to acetyl CoA) proton gradients could drive the origins of both carbon and energy metabolism, leading to the accumulation of organic molecules within vent pores, while facilitating 'dehydration' reactions that form

---

not on the absolute concentrations of ions. Because Na^+ concentration is so high in the oceans, to maintain an equivalent three orders of magnitude difference between the inside and outside of the cell requires pumping a lot more Na^+ than H^+, undermining the advantage of pumping Na^+ if the membrane is relatively impermeable to both ions. Intriguingly, cells that live in vents, such as methanogens and acetogens, often do pump Na^+. One possible reason is that high concentrations of organic acids, such as acetic acid, increase the permeability of the membrane to H^+, making it more profitable to pump Na^+.

complex polymers including DNA, RNA and proteins. I was evasive on details such as how the genetic code arose, but focused on the conceptual argument that these conditions could theoretically have produced rudimentary cells with genes and proteins. Populations of cells were subject to perfectly normal natural selection. I suggested that the last common ancestor of bacteria and archaea, LUCA, may have been the product of selection acting on such populations of simple cells living in the pores of alkaline hydrothermal vents and dependent on natural proton gradients. Selection gave rise to sophisticated proteins, including ribosomes, *Ech* and the ATP synthase – all of which are universally conserved.

In principle, LUCA could have powered all of her carbon and energy metabolism with natural proton gradients, via the ATP synthase and *Ech*, but to do so needed extremely permeable cell membranes. She could not have evolved 'modern' impermeable membranes equivalent to either the bacteria or archaea, because that would have collapsed the natural proton gradients. But an antiporter would have helped, by converting natural proton gradients into biochemical sodium gradients, increasing the power available and so permitting cells to survive on smaller gradients. This would have enabled cells to spread and colonise previously untenable regions of vents, in turn facilitating divergence of populations. Being able to survive under a wider range of conditions could even have enabled cells to 'infect' contiguous vent systems, potentially spreading widely across the ocean floor of the early earth, much of which may have been prone to serpentinisation.

But an antiporter also gave an advantage to pumping, for the first time. Finally we come to those strange differences in the acetyl CoA pathway in methanogens and acetogens. These differences suggest that active pumping arose independently in two distinct populations, which had diverged from a common ancestral population with the help of an antiporter. Recall that methanogens are archaea, while acetogens are bacteria – representatives of the two great domains of prokaryotes, the deepest branches of the 'tree of life'. We have noted that bacteria and archaea are similar in their DNA transcription and translation, ribosomes, protein synthesis, and so on, but differ in other fundamental respects, including cell membrane composition.

I mentioned that they also differ in details of the acetyl CoA pathway, while claiming that this pathway is nonetheless ancestral. The similarities and differences are revealing.

Like methanogens, acetogens react H_2 with CO_2 to form acetyl CoA, via a series of analogous steps. Both groups use a clever trick known as electron bifurcation to power pumping. Electron bifurcation was discovered only recently by the distinguished microbiologist Rolf Thauer and his colleagues in Germany, in what could be the biggest breakthrough in bioenergetics of recent decades. Thauer has now formally retired, but his findings were the culmination of decades spent puzzling over the energetics of obscure microbes, which keep growing when the stoichiometric calculations say they should not. Evolution, as so often, is cleverer than we are. In essence, electron bifurcation amounts to a short-term energy loan, made on the promise of prompt repayment. As we've noted, the reaction of H_2 with CO_2 is exergonic overall (releasing energy) but the first few steps are endergonic (requiring an energy input). Electron bifurcation contrives to use some of the energy that is released in the later, exergonic, steps of CO_2 reduction to pay for the difficult first steps.[7] As more energy is released in the last few

---

7 For those who want to know more about this curious process of electron bifurcation: two separate reactions are coupled together, so that the difficult (endergonic) step is driven by a more favourable (exergonic) reaction. Of the two electrons in H_2, one reacts immediately with an 'easy' target, forcing the other to accomplish a more difficult step, the reduction of CO_2 to organic molecules. The protein machinery that carries out electron bifurcation contains many iron–nickel–sulphur clusters. In methanogens, these essentially mineral structures split up the pairs of electrons from H_2, ultimately feeding half of them on to CO_2, to form organics, and the other half on to sulphur atoms – the 'easier' target that drives the whole process. The electrons are finally reunited on methane (CH_4), which is released into the world as waste, bequeathing methanogens their name. In other words, the process of electron bifurcation is quite staggeringly circular. The electrons from H_2 are separated for a little while, but in the end all of them are transferred on to CO_2, reducing it to methane, which is swiftly discarded. The only thing conserved is some of the energy released in the exergonic steps of CO_2 reduction, in the form of an H^+ gradient across a membrane (actually, in methanogens the gradient is typically Na^+, but H^+ and Na^+ are easily interchangeable via the antiporter). In sum, electron bifurcation pumps protons, regenerating what vents provide for free.

steps than needs to be spent in the first few steps, some energy can be conserved as a proton gradient across a membrane (**Figure 18**). Overall, the energy released by the reaction of H_2 and CO_2 powers the extrusion of protons across a membrane.

The puzzle is that the 'wiring' of electron bifurcation differs in methanogens and acetogens. Both depend on rather similar iron–nickel–sulphur proteins; but the exact mechanism differs, as do many of the proteins needed. Like methanogens, acetogens conserve the energy released by the reaction of H_2 and CO_2 as an H^+ or Na^+ gradient across a membrane. In both cases, the gradient is used to power carbon and energy metabolism. Like methanogens, acetogens have an ATP synthase and *Ech*. Unlike methanogens, however, acetogens do not use the *Ech* to power carbon metabolism directly. On the contrary, some of them use it in reverse as an H^+ or Na^+ pump. And the exact pathway that they use to drive carbon metabolism is very different. These differences seem to be fundamental, to the point that some experts believe the similarities to be the product of convergent evolution or lateral gene transfer, rather than common ancestry.

Yet the similarities and differences begin to make sense if we assume that LUCA did indeed depend on natural proton gradients. If so, the key to pumping could lie in the direction of proton flux through *Ech* – whether the natural flow of protons into the cell drives carbon fixation, or whether this flux is reversed, with the protein now acting as a membrane pump, pumping protons out of the cell (**Figure 20**). In the ancestral population, I suggest that the normal inward flux of protons via *Ech* was used to reduce ferredoxin, in turn driving CO_2 reduction. Two separate populations then invented pumping independently. One population, which ultimately became acetogens, reversed the direction of *Ech*, now oxidising ferredoxin and using the energy released to pump protons out of the cell. This is nice and simple but created an immediate problem. The ferredoxin previously used to reduce carbon is now used to pump protons. Acetogens had to come up with a new way of reducing carbon that did not rely on ferredoxin. Their ancestors found a way – the clever trick of electron bifurcation, which enabled them to reduce CO_2 indirectly. The basic biochemistry of acetogens arguably follows from that simple premise – the direction of proton

A

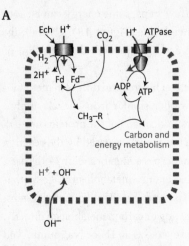

B

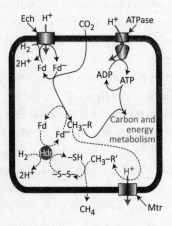

C

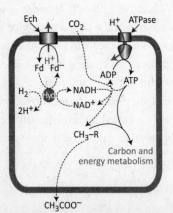

Figure 20 **Possible evolution of active pumping**

Hypothetical origins of pumping in bacteria and archaea, based on the direction of H^+ flux through the membrane protein *Ech*. **A** The ancestral state, in which natural proton gradients drive carbon and energy metabolism via *Ech* and the ATP synthase (ATPase). This can only work so long as the membrane is leaky to protons. **B** Methanogens (postulated to be the ancestral archaea). These cells continue to use *Ech* and the ATPase to drive carbon and energy metabolism, but with H^+-tight membranes could no longer rely on natural proton gradients. They had to 'invent' a new biochemical pathway and new pump (methyl transferase, Mtr) to generate their own H^+ (or Na^+) gradient (dotted lines). Note that this panel is equivalent to Figure 18 A and B combined. **C** Acetogens (postulated to be the ancestral bacteria). The direction of H^+ flux through *Ech* is here reversed, and is now powered by the oxidation of ferredoxin. Acetogens did not need to 'invent' a pump, but had to find a new way of reducing CO_2 to organics; this is done using NADH and ATP (dotted lines). This postulated scenario could explain both the similarities and differences in the acetyl CoA pathway between methanogens and acetogens.

flux through *Ech* was reversed, giving the acetogens a functional pump, but leaving them with a specific set of problems to solve.

The second population, which became methanogens, found an alternative way. Like their ancestors, they continued to use proton gradients to reduce ferredoxin, and then used the reduced ferredoxin to fix carbon. But they then had to 'invent' a pump from scratch. Well, not quite from scratch; they may have repurposed an existing protein. It seems they modified an antiporter to become a straight pump. That is not intrinsically difficult to do, but it gave rise to a different problem: how to power the pump? Methanogens came up with a different form of electron bifurcation, using some of the same proteins as acetogens, but hooked up quite differently, as their own requirements were distinct and wired up to a different pump. The carbon and energy metabolism of each of these domains arguably stems from the direction of proton flow through *Ech*. It's a binary choice, and the methanogens and acetogens made different decisions (**Figure 20**).

Once each group had active pumps, there was finally an advantage to improving the membrane. For all the steps until now, there had never been any benefit to evolving a 'modern' membrane, replete with phospholipids – that would have been actively detrimental. But as soon as cells had antiporters and ion pumps, there was now a benefit to incorporating glycerol head-groups on membrane lipids. And the two domains appear to have done so independently, with archaea using one stereoisomer of glycerol, and bacteria using its mirror image (see Chapter 2).

Now cells had evolved active ion pumps and modern membranes, and were finally free to leave the vents, escaping into the open oceans. From a common ancestor that lived from proton gradients in vents, the first free-living cells, bacteria and archaea, emerged independently. It's not surprising that bacteria and archaea should have come up with distinct cell walls to protect them against these new shocks, nor indeed that they should have 'invented' DNA replication independently. Bacteria attach their DNA to the cell membrane during cell division, at a site called the replicon; the attachment enables each daughter cell to receive a copy of the genome. The molecular machinery required to attach DNA to the membrane, and many details of DNA replication, must depend at least partly on the mechanics of that attachment. The fact that cell membranes evolved independently begins to explain why DNA replication should be so different in bacteria and archaea. Much the same applies to cell walls, all the components of which must be exported from inside the cell through specific membrane pores – hence the synthesis of the cell wall depends on the properties of the membrane, and *should* differ in bacteria and archaea.

And so we draw to a close. While bioenergetics do not predict from first principles that there should be fundamental differences between bacteria and archaea, these considerations do explain how and why they could have arisen in the first place. The deep differences between the prokaryotic domains had nothing to do with adaptation to extreme environments, such as high temperatures, but rather the divergence of cells with membranes that were obliged to remain leaky for bioenergetic reasons. While the divergence of archaea and bacteria might not be predictable from first principles, the fact that both groups are chemiosmotic (depending on proton gradients across membranes) does follow from the physical principles discussed in these last two chapters. The environment most realistically capable of giving rise to life, whether here or anywhere else in the universe, is alkaline hydrothermal vents. Such vents constrain cells to make use of natural proton gradients, and ultimately to generate their own. In this context it's no mystery that all cells here on earth should be chemiosmotic. I would expect that cells across the universe will be chemiosmotic too. And that means they will face exactly the same problems that life on earth does. In the next part, we'll see why this universal requirement for proton power predicts that complex life will be rare in the universe.

PART III

COMPLEXITY

THE ORIGIN OF COMPLEX CELLS

There's a famous line, delivered by Orson Welles, in the 1940s film noir *The Third Man*: 'In Italy, for thirty years under the Borgias, they had warfare, terror, murder and bloodshed, but they produced Michelangelo, Leonardo da Vinci and the Renaissance. In Switzerland, they had brotherly love, they had five hundred years of democracy and peace – and what did that produce? The cuckoo clock.' Welles is said to have written that line himself. The Swiss government reputedly sent him an angry letter, in which they wrote 'We don't make cuckoo clocks.' I don't have anything against the Swiss (or Orson Welles); I tell this story only because, to my own mind, it echoes evolution. Since the first complex eukaryotic cells arose, some 1.5 to 2 billion years ago, we have had warfare, terror, murder and bloodshed: nature, red in tooth and claw. But in the preceding aeons, we had 2 billion years of peace and symbiosis, bacterial love (and not only love), and what did this infinity of prokaryotes come up with? Certainly nothing as large or outwardly complex as a cuckoo clock. In the realm of morphological complexity, neither bacteria nor archaea begin to compare with even single-celled eukaryotes.

It's worth emphasising this point. The two great domains of prokaryotes, the bacteria and archaea, have extraordinary genetic and biochemical versatility. In their metabolism, they put the eukaryotes to shame: a single bacterium can have more metabolic versatility than the entire eukaryotic domain. Yet for some reason, neither bacteria nor archaea ever gave rise *directly* to structural complexity on anything like the eukaryotic scale. In their cell volume, prokaryotes are typically about 15,000 times smaller than eukaryotes (although there are some revealing exceptions, which we'll come to). While there is some overlap in genome size, the largest known bacterial genomes contain about 12 megabases of DNA. In comparison, humans have about 3,000 megabases, and some eukaryotic genomes range up to 100,000 megabases or more. Most compellingly, the bacteria and archaea have barely changed in 4 billion years of evolution. There have been massive environmental upheavals in that time. The rise of oxygen in the air and oceans transformed environmental opportunities, but the bacteria remained unchanged. Glaciations on a global scale (snowball earths) must have pushed ecosystems to the brink of collapse, yet bacteria remained unchanged. The Cambrian explosion conjured up animals – pastures new for bacteria to exploit. Through our human prism, we tend to see bacteria mainly as pathogens, even though the agents of disease are a mere tip of prokaryotic diversity. Yet throughout these shifts, the bacteria remained resolutely bacterial. Never did they give rise to something as large and complex as a flea. Nothing is more conservative than a bacterium.

In Chapter 1, I argued that these facts are best explained in terms of a structural constraint. There is something about the physical structure of eukaryotes that is fundamentally different from both the bacteria and archaea. Overcoming this structural constraint enabled the eukaryotes alone to explore the realm of morphological variation. In the broadest of terms, prokaryotes explored the possibilities of metabolism, finding ingenious solutions to the most arcane chemical challenges, while eukaryotes turned their back on this chemical cleverness, and explored instead the untapped potential of larger size and greater structural complexity.

There is nothing radical about the idea of structural constraints, but of course there is no consensus on what those constraints might be. Many ideas have been put forward, from the catastrophic loss of the cell wall to the novelty of straight chromosomes. Loss of the cell wall can be a catastrophe, as without that rigid external scaffold, cells easily swell and burst. At the same time, however, a straitjacket prevents cells from physically changing their shape, crawling around and engulfing other cells by phagocytosis. A rare successful loss of the cell wall might therefore have permitted the evolution of phagocytosis – an innovation that Oxford biologist Tom Cavalier-Smith has long argued was key to the evolution of eukaryotes. It is true that the loss of the cell wall is necessary for phagocytosis, but many bacteria lose their cell wall and it is often far from catastrophic – so-called L-form bacteria do perfectly well without a cell wall, but show no sign of evolving into dynamic phagocytes. And quite a few archaea do not have a cell wall at all, but likewise do not become phagocytes. To claim that the cumbersome cell wall is *the* constraint that prevented both the bacteria and archaea from evolving greater complexity hardly stands up to scrutiny if many bacteria and archaea lose their cell wall but don't become more complex, whereas many eukaryotes, including plants and fungi, have a cell wall (albeit different from prokaryotic walls) but are nonetheless far more complex than prokaryotes. A telling example is eukaryotic algae compared with cyanobacteria: both have similar lifestyles, living by photosynthesis, both have cell walls; but algal genomes are typically several orders of magnitude larger, encompassing far greater cell volume and structural complexity.

Straight chromosomes suffer from a similar problem. Prokaryotic chromosomes are usually circular, and DNA replication begins at a particular point on that ring (the replicon). However, DNA replication is often slower than cell division, and a cell can't finish dividing in two until it has completed copying its DNA. This means a single replicon limits the maximal size of a bacterial chromosome, because cells with smaller chromosomes will tend to replicate faster than cells with a larger chromosome. If a cell loses any unnecessary genes, it can divide faster. Over time, bacteria with smaller chromosomes will tend to prevail, especially if they can regain any genes they previously lost, but now need again, by lateral gene transfer. In contrast, eukaryotes typically have a number of straight chromosomes, each one of which has multiple replicons. This means that DNA replication operates in parallel in eukaryotes, but in series in bacteria. Yet here again, this constraint hardly explains why prokaryotes could not evolve multiple straight chromosomes; indeed, some bacteria and archaea do now turn out to have straight chromosomes and 'parallel processing', but even so they have not expanded their genome size in the manner of eukaryotes. Something else must be stopping them.

Practically all the structural constraints posited to explain why bacteria don't keep on giving rise to eukaryotic complexity suffer from exactly the same problem: there are plenty of exceptions to each purported 'rule'. As the celebrated evolutionary biologist John Maynard Smith used to say, with crushing politeness, these explanations simply will not do.

So what will do? We have seen that phylogenetics do not offer an easy answer. The last common ancestor of eukaryotes was a complex cell that already had straight chromosomes, a membrane-bound nucleus, mitochondria, various specialised 'organelles' and other membrane structures, a dynamic cytoskeleton, and traits like sex. It was recognisably a 'modern' eukaryotic cell. None of these traits exist in bacteria in anything resembling the eukaryotic state. This phylogenetic 'event horizon' means that the evolution of eukaryotic traits can't be traced back in time beyond the last eukaryotic common ancestor. It's as if every single invention of modern society – houses, hygiene, roads, division of labour, farming, courts of law, standing armies, universities, governments, you name it – all these inventions could be traced back to ancient Rome; but before Rome, there was nothing but primitive hunter-gatherer societies. No remains of ancient Greece, China, Egypt, the Levant, Persia, or any other civilisation; just abundant traces of hunter-gatherers everywhere you look. Here's the rub. Imagine that experts have spent decades scrutinising the archaeology of the world to unearth the remains of earlier cities, civilisations that pre-dated the Romans, which could give some insight into how Rome was built. Hundreds of examples were discovered, yet each one, on closer inspection, turned out to post-date Rome. All these outwardly ancient and primitive cities were actually founded in the 'dark ages' by progenitors who could trace their own ancestors back to ancient Rome. In effect, all roads lead to Rome, and Rome really was built in a day.

That may seem an absurd fantasy, but it is close to the situation that confronts us in biology at the moment. There really are no intermediate 'civilisations' between bacteria and eukaryotes. The few that masquerade as intermediates (the 'archezoa' that we discussed in Chapter 1) were once more glorious, like the shell of Byzantium as the empire shrank in upon the city walls in its last centuries. How can we make head or tail of this scandalous state of affairs? Phylogenetics do, in fact, offer a clue, a clue that necessarily evaded studies of single genes, but has been unmasked in the modern era of full genome comparisons.

The chimeric origin of complexity

The problem with reconstructing evolution from a single gene (even one as highly conserved as the commonly used ribosomal RNA gene) is that, by definition, a single gene produces a branching tree. A single gene cannot have two distinct histories in the same organism – it cannot be chimeric.[1] In an ideal world (for phylogeneticists) each gene would produce a similar tree, reflecting a shared history, but we have seen that this rarely happens in the deep evolutionary past. The usual approach is to fall back on the few genes that do share a history – literally a few dozen at the most – and claim that this is the 'one true phylogenetic tree'. If that were the case, then eukaryotes would be closely related to archaea. This is the standard 'textbook' tree of life (**Figure 15**). Precisely how the eukaryotes relate to archaea is disputed (different methods and genes give different answers) but eukaryotes were for a long time claimed to be a 'sister' group to the archaea. I like to show this standard tree of life when giving lectures. The branch lengths indicate genetic distance. Plainly there is as much gene variation among the bacteria and archaea as there is in eukaryotes – so what happened in that long branch separating the archaea from the eukaryotes? No hint of a clue is hidden in this tree.

Take whole genomes, though, and a completely different pattern emerges. Many eukaryotic genes do not have any equivalents in bacteria or archaea, although this proportion is shrinking as methods become more powerful.

---

1 Actually technically it can, as a single gene can be spliced together from two separate pieces with different histories; but in general this does not happen, and in trying to trace history from single genes, phylogeneticists do not usually set out to reconstruct conflicting stories.

These unique genes are known as eukaryotic 'signature' genes. But even by standard methods, roughly one-third of eukaryotic genes *do* have equivalents in prokaryotes. These genes must share a common ancestor with their prokaryotic cousins; they're said to be homologous. Here's what's interesting. Different genes in the same eukaryotic organism do not all share the same ancestor. Around three-quarters of eukaryotic genes that have prokaryotic homologues apparently have bacterial ancestry, whereas the remaining quarter seem to derive from archaea. That's true of humans, but we are not alone. Yeasts are remarkably similar; so too are fruit flies, sea urchins and cycads. At the level of our genomes, it seems that *all* eukaryotes are monstrous chimeras.

That much is incontestable. What it means is bitterly contested. Eukaryotic 'signature' genes, for example, do not share sequence similarities with prokaryotic genes. Why not? Well, they could be ancient, dating back to the origin of life – what we might call the venerable eukaryote hypothesis. These genes diverged from a common ancestor so long ago that any resemblance has been lost in the mists of time. If that were the case, then eukaryotes must have picked up various prokaryotic genes much more recently, for example when they acquired mitochondria.

This hoary old idea retains an emotional appeal to those who venerate eukaryotes. Emotions and personality play a surprisingly big role in science. Some researchers naturally embrace the idea of abrupt catastrophic changes, whereas others prefer to emphasise continual small modifications – evolution by jerks versus evolution by creeps, as the old joke has it. Both happen. In the case of the eukaryotes, the problem seems to be a case of anthropocentric dignity. We are eukaryotes, and it offends our dignity to see ourselves as Johnny-come-lately genetic mongrels. Some scientists like to view the eukaryotes as descending from the very base of the tree of life, for what I see as basically emotional reasons. It is hard to prove that view wrong; but if it is true, then why did it take so long for eukaryotes to 'take off', to become large and complex? The delay was 2.5 billion years. Why do we see no traces of ancient eukaryotes in the fossil record (despite seeing plenty of prokaryotes)? And if eukaryotes were successful for so long, why are there no surviving early eukaryotes from this extended period before the acquisition of mitochondria? We have seen that there is no reason to suppose that they were outcompeted to extinction, as the existence of the archezoa (see Chapter 1) proves that morphologically simple eukaryotes can survive for possibly hundreds of millions of years alongside bacteria and more complex eukaryotes.

An alternative explanation for the eukaryotic signature genes is simply that they evolved faster than the other genes, and hence lost any former sequence similarity. Why would they evolve so much faster? They would do so if they had been selected for different functions from their prokaryotic ancestors. That sounds entirely reasonable to my ears. We know that eukaryotes have lots of gene families, in which scores of duplicated genes specialise to perform different tasks. Because eukaryotes explored a morphological realm that is barred to prokaryotes, for whatever reason, it is hardly surprising that their genes should have adapted to carry out completely new tasks, losing their erstwhile similarity to their prokaryotic ancestors. The prediction is that these genes do in fact have ancestors among bacterial or archaeal genes, but that adaptation to new tasks expunged their earlier history. I will argue later on that this is indeed the case. For now, let's just note that the existence of eukaryotic 'signature' genes does not exclude the possibility that the eukaryotic cell is fundamentally chimeric – the product of some sort of merger between prokaryotes.

So what about the eukaryotic genes that *do* have identifiable prokaryotic homologues? Why should some of them come from bacteria and some from archaea? This is wholly consistent with a chimeric origin, obviously. The real question concerns the number of sources. Take the 'bacterial' genes in eukaryotes. By comparing whole eukaryotic genomes with bacteria, the pioneering phylogeneticist James McInerney has shown that bacterial genes in eukaryotes are associated with many different bacterial groups. When depicted on a phylogenetic tree, they 'branch' with different groups. By no means do all the bacterial genes found in eukaryotes branch with a single group of modern bacteria such as the α-proteobacteria, as might be supposed if they all derived from the bacterial ancestors of mitochondria. Quite the reverse: at least 25 different groups of modern bacteria appear to have contributed genes to eukaryotes. Much the same goes for archaea, although fewer archaeal groups look to have contributed. What is more curious is that all these bacterial and archaeal genes branch together within the eukaryotic tree, as shown by Bill Martin (**Figure 21**). Plainly they were acquired by the eukaryotes early on in evolution, and have shared a common history ever since. That rules out a steady flow of lateral gene transfer over the whole course of eukaryotic history. Something odd seems to have happened at the very origin of eukaryotes. It looks like the first eukaryotes picked up thousands of genes from prokaryotes, but then ceased to ply any trade in prokaryotic genes. The simplest explanation for this picture is not bacterial-style lateral gene transfer, but eukaryotic-style endosymbiosis.

On the face of it, there could have been scores of endosymbioses, as indeed predicted by the serial endosymbiosis theory. Yet it is barely credible that there could have been 25 different bacteria and 7 or 8 archaea all contributing to an early orgy of endosymbioses, a cellular love-fest; and then nothing for the rest of eukaryotic history. But if not that, then what else could explain this pattern? There is a very simple explanation – lateral gene transfer. I'm not contradicting myself. There could have been a single endosymbiosis at the origin of eukaryotes, and then almost no further exchange of genes between bacteria and eukaryotes; but plenty of lateral gene transfer over the entire period between various groups of bacteria. Why would eukaryotic genes branch with 25 different groups of bacteria? They would do so if eukaryotes acquired a large number of genes from a single population of bacteria – a population that subsequently changed over time. Take a random assortment of genes from the 25 different groups of bacteria and place them all together in a single population. Let's say that these bacteria were the ancestors of mitochondria, and they lived around 1.5 billion years ago. There are no cells quite like them today, but given the prevalence of lateral gene transfer in bacteria, why should there be? Some of this population of bacteria were acquired by endosymbiosis, whereas others retained their freedom as bacteria, and spent the next 1.5 billion years swapping their genes by lateral transfer, as modern bacteria do. So the ancestral hand of genes was dealt across scores of modern groups.

The same goes for the host cell. Take the genes from the 7 or 8 groups of archaea that contributed to eukaryotes, and place them in an ancestral

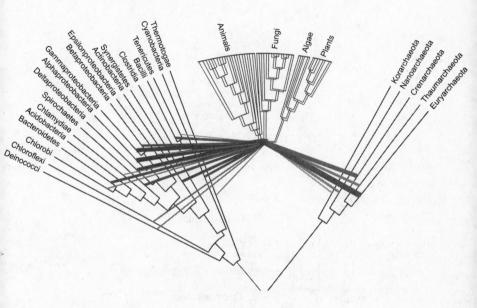

Figure 21 **The remarkable chimerism of eukaryotes**
Many eukaryotic genes have equivalents in bacteria or archaea, but the range of apparent sources is startling, as seen in this tree by Bill Martin and colleagues. The tree depicts the closest matches to specific bacterial or archaeal groups for eukaryotic genes with clear prokaryotic ancestry. Thicker lines indicate that more genes apparently derive from that source. For example, a large proportion of genes appears to derive from the Euryarchaeota. The range of sources could be interpreted as multiple endosymbioses or lateral gene transfers, but there is no morphological evidence for this, and it is difficult to explain why all of these prokaryotic genes branch together within the eukaryotes; that implies there was a short evolutionary window early in eukaryotic evolution when genetic transfers were rife, followed by next to nothing for the following 1.5 billion years. A simpler and more realistic explanation is that there was a single endosymbiosis between an archaeon and a bacterium, neither of which had a genome equivalent to any modern group; and subsequent lateral gene between the descendants of these cells and other prokaryotes gave rise to modern groups with an assortment of genes.

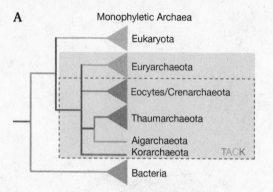

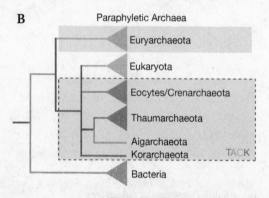

Figure 22 **Two, not three, primary domains of life**
Seminal work by Martin Embley and colleagues shows that eukaryotes derive from archaea, reproduced courtesy of Nature Publishing Group. **A** shows a conventional three-domains tree, in which each domain is monophyletic (unmixed): the eukaryotes are at the top, the bacteria at the bottom, and the archaea are shown split into several large groups that are more closely related to each other than to either the bacteria or eukaryotes. **B** shows a more recent and strongly supported alternative tree, based on far wider sampling and a larger number of informational genes involved in transcription and translation. The informational genes of eukaryotes here branch *within* the archaea, close to a specific group known as eocytes, hence the name of the hypothesis. The implication is that the host cell that acquired a bacterial endosymbiont at the origin of the eukaryotic domain was a *bona fide* archaeon, something like an eocyte, and was therefore not some sort of 'primitive phagocyte'. TACK stands for the superpylum comprising Thaumarchaeota, Aigarcheota, Crenarchaeota and Korarchaeota.

population that lived 1.5 billion years ago. Again, some of these cells acquired endosymbionts – which ultimately evolved into mitochondria – while the rest just kept doing what archaea do, swapping genes around by lateral gene transfer. Notice that this scenario is reverse engineering, and assumes no more than what we already know to be true: that lateral gene transfer is common in bacteria and archaea, and much less common in eukaryotes. It also assumes that one prokaryote (an archaeon, which by definition is not capable of engulfing other cells by phagocytosis) could acquire endosymbionts by some other mechanism. We'll put that aside for now and return to it later.

This is the simplest possible scenario for the origin of eukaryotes: there was a single chimeric event between an archaeal host cell and a bacterial endosymbiont. I do not expect you to believe me at this point. I am arguing simply that this scenario is compatible with everything we know about the phylogenetic history of eukaryotes, as are several other possible scenarios. I favour this view on the basis of Occam's razor alone (it is the simplest explanation of the data), though there is increasingly powerful phylogenetic evidence from Martin Embley and colleagues in Newcastle that this is exactly what happened (**Figure 22**). But given that eukaryotic phylogenetics remains controversial, can the question be resolved in some other way? I think so. If the eukaryotes arose in an endosymbiosis between two prokaryotes, an archaeal host cell and a bacterial endosymbiont, which went on to become mitochondria, then we can explore the question from a more conceptual point of view. Can we think of a good reason why one cell getting inside another cell should transform the prospects of prokaryotes, unleashing the potential of eukaryotic complexity? Yes. There is a compelling reason, and it relates to energy.

Why bacteria are still bacteria

The key to it all is that prokaryotes – both bacteria and archaea – are chemiosmotic. We saw in the previous chapter how the first cells might have arisen within the rocky walls of hydrothermal vents, how natural proton gradients could have driven both carbon and energy metabolism, and why this reliance on proton gradients could have forced the deep split between bacteria and archaea. These considerations could indeed explain how chemiosmotic coupling first arose, but they do not explain why it persisted for evermore in all bacteria, all archaea, and all eukaryotes. Was it not possible for some groups to lose chemiosmotic coupling, to replace it with something else, something better?

Some groups did. Yeasts, for example, spend much of their time fermenting, as do a few bacteria. The process of fermentation generates energy in the form of ATP, but although faster, fermentation is an inefficient use of resources. Strict fermenters soon pollute their environment, preventing themselves from growing, while their wasteful end products, such as ethanol or lactate, are fuels for other cells. Chemiosmotic cells can burn these waste products with oxygen or other substances, such as nitrate, to glean far more energy, permitting them to keep on growing for longer. Fermentation works well as part of a mix in which other cells burn up the end products, but is very limited by itself.[2] There is strong evidence that fermentation arose later in evolution than respiration, and that makes perfect sense in light of these thermodynamic limitations.

Perhaps surprisingly, fermentation is the only known alternative to chemiosmotic coupling. All forms of respiration, all forms of photosynthesis, and indeed all forms of autotrophy, where cells grow from simple inorganic precursors only, are strictly chemiosmotic. We noted some good reasons for this in Chapter 2. In particular, chemiosmotic coupling is marvellously versatile. A massive range of electron sources and sinks can all be plugged into a common operating system, allowing small adaptations to have an immediate benefit. Likewise, genes can be passed around by lateral gene transfer, and again can be installed into a fully compatible system, like a new app. So chemiosmotic coupling enables metabolic adaptation to almost any environment in almost no time at all. No wonder it dominates!

---

2 Much the fastest and most reliable way of removing the end products of fermentation is burning them, via respiration. The end product, CO_2, is lost simply by diffusion into the air, or precipitation as carbonate rocks. Fermentation therefore depends largely on respiration.

But that's not all. Chemiosmotic coupling also enables the last drops of energy to be squeezed from any environment. Take methanogens, which use H_2 and CO_2 to drive carbon and energy metabolism. We have noted that it's not easy to get H_2 and CO_2 to react together: an input of energy is needed to overcome the barrier to their reaction; methanogens use that clever trick called electron bifurcation to coerce them into reacting. In terms of the overall energetics, think of the Hindenburg airship, the German dirigible filled with hydrogen gas, which erupted like a firebomb after crossing the Atlantic, giving hydrogen a bad name ever since. H_2 and O_2 are stable and unreactive so long as energy is not added in the form of a spark. Even a small spark immediately releases a vast amount of energy. In the case of H_2 and CO_2, the problem is reversed – the 'spark' has to be relatively large, whereas the amount of energy released is rather small.

Cells face an interesting limitation if the amount of usable energy released from any one reaction is less than double the energy input required. You may recall having to balance chemical equations at school. A whole molecule must react with another molecule – it's impossible for half a molecule to react with three-quarters of another one. For a cell, 1 ATP must be spent to gain fewer than 2 ATPs. There is no such thing as 1.5 ATPs – there can be either one or two. So 1 ATP has to be spent to gain 1 ATP. There is no net gain, and that precludes growth from H_2 and CO_2 by normal chemistry. This is true not only of H_2 and CO_2 but also of many other redox couples (a pairing of electron donor and acceptor), such as methane and sulphate. Despite this basic limitation of chemistry, cells still grow from these redox couples perfectly happily. They do so because proton gradients across membranes are by definition *gradations*. The beauty of chemiosmotic coupling is that it transcends chemistry. It allows cells to save up 'loose change'. If it takes 10 protons to make 1 ATP, and a particular chemical reaction only releases enough energy to pump 4 protons, then the reaction can simply be repeated 3 times to pump 12 protons, 10 of which are then used to make 1 ATP. While this is strictly necessary for some forms of respiration, it is beneficial for all of us, as it allows cells to conserve small amounts of energy that would otherwise be wasted as heat. And that, almost always, gives proton gradients an edge over plain chemistry – the power of nuance.

The energetic benefits of chemiosmotic coupling suffice to explain why it has persisted for 4 billion years; but proton gradients also have other facets that have become incorporated into the function of cells. The more deeply rooted a mechanism, the more it can become the basis of quite unrelated traits. So proton gradients are widely used to drive the uptake of nutrients and excretion of waste; they are used to turn the screw that is the bacterial flagellum, a rotating propeller that motors cells about; and they are deliberately dissipated to produce heat, as in brown fat cells. Most intriguingly, their collapse ushers in the abrupt programmed death of bacterial populations. In essence, when a bacterial cell becomes infected with a virus, it is most likely doomed. If it can kill itself quickly, before the virus copies itself, then its kin (nearby cells sharing related genes) might survive. The genes that orchestrate cell death will spread through the population. But these death genes have to act quickly, and few mechanisms are faster than perforating the cell membrane. Many cells do exactly this – when infected, they form pores in the membrane. These collapse the proton-motive force, which in turn trips the latent death machinery. Proton gradients have become the ultimate sensors of cellular health, the arbiters of life and death, a role that will loom large later in this chapter.

All in all, the universality of chemiosmotic coupling does not look like a fluke. Its origin was arguably linked to the origin of life and the emergence of cells in alkaline hydrothermal vents (by far the most probable incubators of life), while its persistence in almost all cells makes very good sense. What once seemed to be a peculiar mechanism now looks to be only superficially counterintuitive – our analysis suggests that chemiosmotic coupling ought to be literally a universal property of life in the cosmos. And that means that life elsewhere should face exactly the same problem that bacteria and archaea face here, rooted in the fact that prokaryotes pump protons across their cell membrane. That doesn't constrain real prokaryotes in any way – quite the contrary – but it does set limits on what is possible. What is not possible, I will argue, is precisely what we do not see: large morphologically complex prokaryotes with big genomes.

The issue is energy availability per gene. I had been stumbling blindly towards this concept for some years, but it was the cut and thrust of dialogue with Bill Martin that really brought matters to a head. After weeks of talking, trading ideas and perspectives, it suddenly dawned on us that the key to the evolution of eukaryotes lies in the simple idea of 'energy per gene'. With overflowing excitement I spent a week scribbling calculations on the back of an envelope, in the end lots of envelopes, and finally came up with an answer that shocked us both, an answer that extrapolated from data in the literature to put a number on the energy gap that separates prokaryotes from eukaryotes. By our calculations, eukaryotes have up to 200,000 times more energy per gene than prokaryotes. Two hundred thousand times more energy! At last we had a gulf between the two groups, a chasm that explains with visceral force why the bacteria and archaea never evolved into complex eukaryotes, and by the same token, why we are unlikely ever to meet an alien composed of bacterial cells. Imagine being trapped in an energy landscape, where the peaks are high energy, and the troughs low energy. Bacteria sit at the bottom of the deepest trough, in an energy chasm so profound that the walls above stretch high into the sky, utterly unscalable. No wonder prokaryotes remained there for an eternity. Let me explain.

Energy per gene

By and large, scientists compare like with like. When it comes to energy the fairest comparison is per gram. We can compare the metabolic rate of 1 gram of bacteria (measured as oxygen consumption) with 1 gram of eukaryotic cells. I doubt that it will surprise you to learn that bacteria usually respire faster than single-celled eukaryotes, on average three times faster. That unsurprising fact is where most researchers tend to leave it; to go on risks comparing apples with pears. We went on. What if we compared the metabolic rate per cell? What an unfair comparison! In our sampling of about 50 bacterial species and 20 single-celled eukaryotic species, the eukaryotes were (on average) 15,000 times larger than the bacteria in their cell volume.[3] Given that they respire at a third of the bacterial rate, the average

---

3 For these comparisons, we need to know the metabolic rate of each of these cells,

eukaryote consumes about 5,000 times more oxygen per second than the average bacterium. That simply reflects the fact that the eukaryote is much bigger, with far more DNA. Nonetheless, a single eukaryotic cell still has 5,000 times more energy. What is it spending it on?

Not much of this extra energy is spent on DNA itself; only about 2% of the overall energy budget of a single-celled organism goes on replicating DNA. In contrast, according to Frank Harold, distinguished elder states-man of microbial bioenergetics (and a hero of mine, even though we don't always agree), cells spend as much as 80% of their total energy budget on protein synthesis. That's because cells are mostly made of proteins; about half the dry weight of a bacterium is protein. Proteins are also very costly to make – they are strings of amino acids, usually a few hundred of them linked together in a long chain by 'peptide' bonds. Each peptide bond requires at least 5 ATPs to seal, five times as much as is needed to polymerise nucleotides into DNA. And then each protein is produced in thousands of copies, which are continuously turned over to repair wear and tear. To a first approximation, then, the energy costs of cells equate closely to the costs of making proteins. Each different protein is encoded by a single gene. Assuming that all genes are translated into proteins (which is generally the case, despite differences in gene expression), the more genes there are in a genome, the higher the costs of protein synthesis. This is borne out by the simple expedient of counting ribosomes (the protein-building factories in cells), as there is a straight correlation between ribosome number and the burden of protein synthesis. There are about 13,000 ribosomes in an average bacterium such as *E. coli*; and at least 13 million in a single liver cell, about 1,000 to 10,000 times as many.

On average, bacteria have around 5,000 genes, eukaryotes have about

---

as well as cell volume and genome size. If you think that 50 bacteria and 20 eukaryotes is not many for a comparison of this kind, just think of the difficulties involved in procuring all this information for each cell type. There are plenty of cases where the metabolic rate has been measured, but not the genome size or cell volume, or vice versa. Even so, the values that we extracted from the literature seem to be reasonably robust. If you are interested in the detailed calculations, see Lane and Martin (2010).

20,000, ranging up to 40,000 in the case of large protozoa, like the familiar pond-dwelling paramecium (which has twice as many genes as we do). The average eukaryote has 1,200 times as much energy per gene as the average prokaryote. If we correct for the number of genes by scaling up the bacterial genome of 5,000 genes to a eukaryote-sized genome of 20,000 genes, the bacterial energy-per-gene falls to nearly 5,000 times less than the average eukaryote. In other words, eukaryotes can support a genome 5,000 times larger than bacteria, or alternatively, they could spend 5,000 times more ATP on expressing each gene, for example by producing many more copies of each protein; or a mixture of the two, which is in fact the case.

Big deal, I hear you say, the eukaryote is 15,000 times larger. It has to fill up this larger volume with something, and that something is mostly protein. These comparisons only make sense if we correct for cell volume too. Let's expand our bacterium up to the average size for eukaryotes, and calculate how much energy it would have to spend per gene then. You might think a larger bacterium would have more ATP, and indeed it does; but it also has a greater demand for protein synthesis, and that consumes more ATP. The overall balance depends on how those factors interrelate. We calculated that bacteria actually pay a hefty penalty for being bigger: size does matter, and for bacteria, bigger is not better. On the contrary, giant bacteria should have 200,000 times *less* energy per gene than a eukaryote of the same size. Here's why.

Scaling up a bacterium over orders of magnitude immediately runs into a problem with the surface-area-to-volume ratio. Our eukaryote has a mean volume that is 15,000 times larger than an average bacterium. Let's keep things simple, and assume that cells are just spheres. To inflate our bacterium up to eukaryotic size, the radius would need to increase 25-fold, and the surface area 625-fold.[4] This matters, as ATP synthesis takes place across

---

4 The volume of a sphere varies with the cube of its radius, whereas the surface area varies with the square of the radius. Increasing the radius of the sphere therefore increases the volume faster than the surface area, giving cells a problem in that their surface area becomes proportionately smaller in relation to their volume. Changing shape helps: for example, many bacteria are rod-shaped, giving them a larger surface area

the cell membrane. To a first approximation, then, ATP synthesis would increase 625-fold, in line with the expanded membrane surface area.

But of course ATP synthesis requires proteins: respiratory chains that actively pump protons across the membrane, and the ATP synthase, the molecular turbines that use the flow of protons to power ATP synthesis. If the surface area of the membrane is increased 625-fold, ATP synthesis could only expand 625-fold if the total number of respiratory chains and ATP synthase enzymes were increased commensurately, such that their concentration remained the same per unit area. That is surely true, but the reasoning is pernicious. All of these extra proteins need to be physically made and inserted into the membrane, and that requires ribosomes and all kinds of assembly factors. These have to be synthesised too. Amino acids must be delivered to the ribosomes, along with RNAs, all of which have to be made as well, necessitating in turn the genes and proteins needed to do so. To support this extra activity, more nutrients must be shipped across the expanded membrane area, and this requires specific transport proteins. Indeed we need to synthesise the new membrane too, demanding the enzymes for lipid synthesis. And so on. This great tide of activity could not be supported by a single genome. Picture it, one diminutive genome, sitting there all by itself, responsible for producing 625 times as many ribosomes, proteins, RNAs and lipids, somehow shipping them across the vastly expanded cell surface, for what? Merely to sustain ATP synthesis at the same rate, per unit surface area, as before. Plainly that is not possible. Imagine increasing the size of a city 625-fold, with new schools, hospitals, shops, playgrounds, recycling centres, and so on; the local government responsible for all these amenities can hardly be run on the same shoestring.

Given the speed of bacterial growth, and the benefits that accrue from streamlining their genomes, the likelihood is that protein synthesis from each genome is already stretched quite close to its limit. Increasing overall protein synthesis by 625-fold would most reasonably require 625 copies of

---

in relation to their volume; but when expanded in size over several orders of magnitude, such shape changes only mitigate the problem to a degree.

the full bacterial genome to cope, with each genome operating in exactly the same way.

On the face of it, that might sound crazy. In fact, it's not; we'll return to this point in a moment. For now, though, let's just consider the energy costs. We have 625 times as much ATP, but 625 times as many genomes, each one of which has equivalent running costs. In the absence of a sophisticated intracellular transport system, which would take many generations and buckets of energy to evolve, each of these genomes is responsible for an equivalent 'bacterial' volume of cytoplasm, membrane, and so on. Probably the best way to see this scaled-up bacterium is not as a single cell at all, but as a consortium of 625 identical cells, merged into a whole. Plainly the 'energy per gene' remains exactly the same for each of these merged units. Scaling up the surface area of a bacterium therefore has no energy benefit at all. Scaled-up bacteria remain at a major disadvantage compared with eukaryotes. Remember that eukaryotes have 5,000 times more energy per gene than 'normal' bacteria. If scaling up the bacterial surface area by 625-fold has no effect on the energy availability per gene, then that remains 5,000 times lower than in eukaryotes.

It gets worse. We have scaled up the surface area of our cell 625-fold, by multiplying the energetic costs and benefits of bacteria 625 times. But what about the internal volume? That is increased by a whopping 15,000-fold. Our scaling so far has produced a giant bubble of a cell, with an interior that is undefined in metabolic terms; we have left it with zero energy requirements. That would be true if the inside was filled with a giant vacuole, metabolically inert. But if that were the case, our scaled-up bacterium would not compare with a eukaryote, which is not just 15,000 times larger, but is stuffed with complicated biochemical machinery. That's mostly made of proteins too, with similar energetic costs. The same arguments apply if we take all these proteins into account. It is inconceivable that cell volume could be increased 15,000-fold without raising the total number of genomes by roughly the same amount. But ATP synthesis cannot be increased commensurately – it depends on the cell membrane area, and we've already taken that into consideration. So scaling up a bacterium to the size of an average eukaryote increases ATP synthesis by 625-fold, but increases the energy costs by up to 15,000-fold. The energy available per single copy of each gene must fall 25-fold. Multiply that by the 5,000-fold difference in energy per gene (after correcting for genome size), and we see that equalising for both genome size and cell volume means that giant bacteria have 125,000 times less energy per gene than eukaryotes. That's an average eukaryote. Large eukaryotes such as amoebae have more than 200,000 times the energy per gene than a giant scaled-up bacterium. That's where our number came from.

You might think this is just playing trivial games with numbers, that it holds no real meaning. I must confess that worried me too – these numbers are quite literally incredible – but this theorising does at least make a clear prediction. Giant bacteria should have thousands of copies of their full genome. Well, that prediction is easily testable. There are some giant bacteria out there; they're not common, but they do exist. Two species have been studied in detail. *Epulopiscium* is known only from the anaerobic hind gut of surgeonfish. It is a battleship of a cell – long and streamlined, about half a millimetre in length, just visible to the naked eye. That's substantially larger than most eukaryotes, including paramecium (**Figure 23**). Why *Epulopiscium* is so big is unknown. *Thiomargarita* is even larger. These cells are spheres, nearly a millimetre in diameter and composed mostly of a huge vacuole. A single cell can be as big as the head of a fruit fly! *Thiomargarita* lives in ocean waters periodically enriched in nitrates by upwelling currents. The cells trap the nitrates in their vacuoles for use as electron acceptors in respiration, enabling them to keep respiring during days or weeks of nitrate deprivation. But that is not the point. The point is that both *Epulopiscium* and *Thiomargarita* exhibit 'extreme polyploidy'. That means they have thousands of copies of their full genome – up to 200,000 copies in the case of *Epulopiscium* and 18,000 copies in the case of *Thiomargarita* (despite most of the cell being a huge vacuole).

Suddenly, loose talk of 15,000 genomes doesn't sound very crazy after all. Not only the number but the distribution of these genomes corresponds to theory. In both cases, they are positioned close to the cell membrane, around the periphery of the cell (**Figure 23**). The centre is metabolically inert, just vacuole in the case of *Thiomargarita*, and a nearly empty

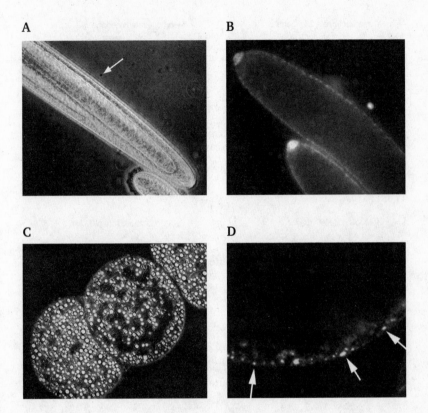

Figure 23 **Giant bacteria with 'extreme polyploidy'**
A shows the giant bacterium *Epulopiscium*. The arrow points to the 'typical' bacterium *E. coli*, for comparison's sake. The cell at the bottom centre is the eukaryotic protist *Paramecium*, dwarfed by this battleship of a bacterium. **B** shows *Epulopiscium* stained by DAPI staining for DNA. The white dots close to the cell membrane are copies of the complete genome – as many as 200,000 copies in larger cells, a state known as 'extreme polyploidy'. **C** is an even larger bacterium, *Thiomargarita*, which is about 0.6 mm in diameter. **D** shows *Thiomargarita* stained by DAPI staining for DNA. Most of the cell is taken up with a giant vacuole, the black area at the top of the micrograph. Surrounding the vacuole is a thin film of cytoplasm containing as many as 20,000 copies of the complete genome (marked by white arrows).

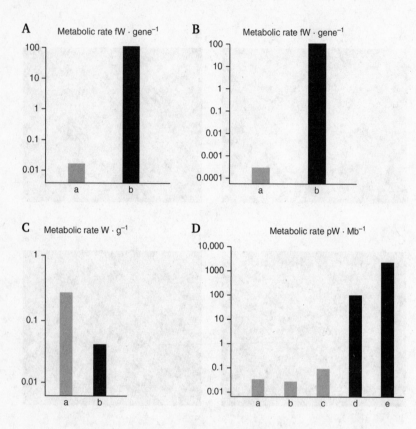

Figure 24 **Energy per gene in bacteria and eukaryotes**
A shows the average metabolic rate per gene in bacteria (a, grey bar) compared
with single-celled eukaryotes (b, black bar), when equalised for genome size.
B shows much the same thing, but this time equalised for cell volume (15,000-fold
larger in eukaryotes) as well as genome size. Notice that the Y axis on all these
graphs is logarithmic, so each unit is a 10-fold increase. A single eukaryotic cell
therefore has 100,000-fold more energy per gene than bacteria, despite respiring
about three times slower per gram of cells (as shown in **C**). These numbers are
based on measured metabolic rates, but corrections for genome size and cell
volume are theoretical. **D** shows that the theory matches reality nicely. The values
shown are the metabolic rate for each single genome, taking into consideration
genome size, copy number (polyploidy) and cell volume. In this case a is *E. coli*, b
is *Thiomargarita*, c is *Epulopiscium*, d is *Euglena*, and e is the large *Amoeba proteus*.

spawning ground for new daughter cells in the case of *Epulopiscium*. The fact that the interior is metabolically almost inert means they save on the costs of protein synthesis, and so don't hoard more genomes inside their insides. In theory, that means they should be roughly comparable with normal bacteria in their energy per gene – the extra genomes are each associated with more bioenergetic membrane, capable of generating all the extra ATP needed to support the additional copies of each gene.

And so, it seems, they are. The metabolic rates of these bacteria have been skilfully measured and we know the total copy number of the genome, so we can calculate the energy per gene directly. And lo! It is close (within the same order of magnitude) to that of the bog-standard bacterium *E. coli*. Whatever the costs and benefits of greater size in giant bacteria may be, there is no energy advantage. Exactly as predicted, these bacteria have about 5,000 times less energy per single copy of each gene than eukaryotes (**Figure 24**). Note that this figure is not 200,000 times less, as these giant bacteria only have multiple genomes around their periphery, and not inside – their inner volume is metabolically nearly inert, giving the giants a problem with cell division, which helps to explain why they're not abundant.

Bacteria and archaea are happy as they are. Small bacteria with small genomes are not energetically limited. The problem only emerges when we try to scale up bacteria to eukaryotic sizes. Rather than swelling up their genome size and energy availability in eukaryotic fashion, energy per gene actually falls. The gulf becomes enormous. Bacteria can't expand their genome size, nor can they accumulate the thousands of new gene families, encoding all kinds of new functions, that epitomise eukaryotes. Rather than evolving a single gigantic nuclear genome, they end up hoarding thousands of copies of their standard-issue small bacterial genome.

How eukaryotes escaped

Why don't the same problems of scale prevent eukaryotes from becoming complex? The difference lies in the mitochondria. Recall that eukaryotes arguably originated in a genomic chimera between an archaeal host cell and a bacterial endosymbiont. The phylogenetic evidence, I said, is consistent with this scenario, but that in itself is not sufficient to prove it. Yet the severe energetic constraints on bacteria come very close to proving a *requirement* for a chimeric origin of complex life. Only an endosymbiosis between prokaryotes, I will argue, could break the energetic constraints on bacteria and archaea – and endosymbioses between prokaryotes are extremely rare in evolution.

Bacteria are autonomous *self*-replicating entities – cells – whereas genomes are not. The problem that faces giant bacteria is that, to be large, they must replicate their whole genome thousands of times. Each genome is copied perfectly, or nearly perfectly, but then it just sits there, unable to do anything else. Proteins may set to work on it, transcribing and translating genes; the host cell may divide, powered by the dynamism of its proteins and metabolism, but the genome itself is wholly inert, as incapable of replicating itself as the hard disk of a computer.

What difference does that make? It means that all the genomes in the cell are essentially identical copies of each other. Differences between them are not subject to natural selection, because they are not self-replicating entities. Any variations between different genomes in the same cell will even out over generations, as so much noise. But consider what happens when whole bacteria compete among themselves. If one line of cells happens to replicate twice as fast as another, it will double its advantage each generation, growing exponentially faster. In just a few generations, the fast-growing line will dominate the population. Such a massive advantage in growth rate might be unlikely, but bacteria grow so quickly that even small differences in growth rate can have a marked effect on the composition of a population over many generations. For bacteria, one day could see the passage of 70 generations, the dawn of that day as remote as the birth of Christ when measured in human lives. Tiny differences in growth rate can be achieved by small deletions of DNA from a genome, such as a loss of one gene that is no longer in use. No matter whether this gene might be needed again in the future, the cells that lose it will replicate a little faster, and within a few days will come to dominate the population. Those that retain the useless gene will slowly be displaced.

Then conditions change again. The useless gene regains its value. Cells lacking it can no longer grow, unless they reacquire it by lateral gene transfer. This endlessly circular dynamic of gene loss and gain dominates bacterial populations. Over time, genome size stabilises at the smallest feasible size, while individual cells have access to a much larger 'metagenome' (the total pool of genes within the whole population, and indeed neighbouring populations). A single *E. coli* cell may have 4,000 genes, but the metagenome is more like 18,000 genes. Dipping into this metagenome carries risks – picking up the wrong gene, or a mutated version, or a genetic parasite instead; but over time the strategy pays off, as natural selection eliminates the less fit cells and the lucky winners take all.

But now think about a population of bacterial endosymbionts. The same general principles apply – this is just another population of bacteria, albeit a small population in a restricted space. Bacteria that lose unnecessary genes will replicate slightly faster and tend to dominate, just as before. The key difference is the stability of the environment. Unlike the great outdoors, where the conditions are always changing, the cytoplasm of cells is a very stable environment. It may not be easy to get there, or to survive there, but once established, a steady and invariable supply of nutrients can be relied upon. The endlessly circular dynamic of gene loss and gain in free-living bacteria is replaced with a trajectory towards gene loss and genetic streamlining. Genes that are not needed will never be needed again. They can be lost for good. Genomes shrink.

I mentioned that endosymbioses are rare between prokaryotes, which are not capable of engulfing other cells by phagocytosis. We do know of a couple of examples in bacteria (**Figure 25**), so plainly they can occur, if only very occasionally in the absence of phagocytosis. A few fungi are also known to have endosymbionts, despite being no more phagocytic than bacteria. But phagocytic eukaryotes frequently have endosymbionts; hundreds of examples are known.[5] They share a common trajectory towards gene

---

5 The fact that prokaryotes cannot engulf other cells by phagocytosis is sometimes cited as a reason why the host cell 'had' to be some kind of 'primitive' phagocyte, not a prokaryote. There are two problems with this reasoning. The first is that it is just not true – we know of rare examples of endosymbionts living within prokaryotes. The

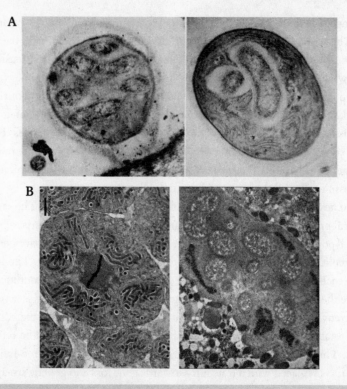

Figure 25 **Bacteria living within other bacteria**
A A population of intracellular bacteria living inside cyanobacteria. The wavy internal membranes in the right-hand cell are thylakoid membranes, the site of photosynthesis in cyanobacteria. The cell wall is the darker line enclosing the cell, which is sheathed within a transluscent gelatinous coat. The intracellular bacteria are enclosed in a lighter space that could be mistaken for a phagocytic vacuole, but is probably an artefact of shrinkage, as no cells with a wall can engulf other cells by phagocytosis. How these bacteria got inside is a mystery, but there's no doubt that they are really there, and so no doubt that it is possible, if very rare, to have intracellular bacteria inside free-living bacteria. **B** Populations of gamma-proteobacteria inside beta-proteobacterial host cells, in turn living within the eukaryotic cells of a multicellular mealybug. On the left, the central cell (with the nucleus about to divide by mitosis) has six bacterial endosymbionts, each of them containing a number of rod-shaped bacteria, shown magnified on the right. This case is less compelling than the cyanobacterial example, as their cohabitation within a eukaryotic cell is not equivalent to a free-living host cell; nonetheless, both cases show that phagocytosis is not needed for an endosymbiosis between bacteria.

loss. The smallest bacterial genomes are usually found in endosymbionts. *Rickettsia*, for example, the cause of typhus and scourge of Napoleon's army, has a genome size of just over 1 megabase, barely a quarter the size of *E. coli*. *Carsonella*, an endosymbiont of jumping plant lice, has the smallest known bacterial genome, which at 200 kilobases is smaller than some plant mitochondrial genomes. Although we know next to nothing about gene loss in endosymbionts within prokaryotes, there is no reason to suppose they would behave any differently. Indeed, we can be sure they would have lost genes in much the same way: mitochondria, after all, were once endosymbionts living in an archaeal host.

Gene loss makes a huge difference. Losing genes is beneficial to the endosymbiont, as it speeds up their replication; but losing genes also saves ATP. Consider this simple thought experiment. Imagine that a host cell has 100 endosymbionts. Each endosymbiont starts out as a normal bacterium, and it loses genes. Let's say it starts out with a fairly standard bacterial genome of 4,000 genes, and it loses 200 of them (5%), perhaps initially the genes for cell-wall synthesis, which are no longer needed when living in a host cell. Each of these 200 genes encodes a protein, which has an energetic cost to synthesise. What are the energy savings of *not* making those proteins? An average bacterial protein has 250 amino acids, with an average of 2,000 copies of each protein. Each peptide bond (which joins amino acids together) costs about 5 ATPs. So the total ATP cost of 2,000 copies of 200 proteins in 100 endosymbionts is 50 billion ATPs. If this energy cost is

---

second problem is precisely that endosymbionts are common in eukaryotes, and yet do not routinely give rise to organelles like mitochondria. Indeed, the only examples known are mitochondria and chloroplasts, despite (no doubt) thousands or millions of opportunities. The origin of the eukaryotic cell was a singular event. As noted in Chapter 1, a proper explanation must elucidate why it happened only once: it must be persuasive enough to be believable, but not so persuasive that we are left wondering why it did not happen on multiple occasions. Endosymbiosis between prokaryotes is rare, but not so rare that it can account for the singularity of eukaryotic origins by itself. However, the enormous energetic rewards of endosymbiosis between prokaryotes, when combined with grave difficulties of reconciling life cycles (which we will discuss in the next chapter), do together explain this evolutionary singularity.

incurred during the life cycle of a cell, and the cell divides every 24 hours, then the cost for synthesising these proteins would be 580,000 ATPs per second! Conversely, that is the ATP saving if those proteins are not made.

There is no necessary reason for these ATPs to be spent on anything else of course (although there are some possible reasons, to which we will return), but let's just consider what kind of difference it could make to a cell if they *were* spent. One relatively simple factor that sets eukaryotes apart from bacteria is a dynamic internal cytoskeleton, capable of remodelling itself and changing shape in the course of either cell movement or the transport of materials within the cell. A major component of the eukaryotic cytoskeleton is a protein called actin. How much actin could we make for 580,000 ATPs per second? Actin is a filament composed of monomers joined together in a chain; and two such chains are wound around each other to form the filament. Each monomer has 374 amino acids, and there are 2 × 29 monomers per micrometre of actin filament. With the same ATP cost per peptide bond, the total ATP requirement per micrometre of actin is 131,000. So in principle we could make about 4.5 micrometres of actin per second. If that doesn't sound much to you, bear in mind that bacteria are typically a couple of micrometres in length.[6] So the energy savings accruing from endosymbiotic gene loss (just 5% of their genes) could easily support the evolution of a dynamic cytoskeleton, as indeed happened. Bear in mind, as well, that 100 endosymbionts is a conservative estimate. Some large amoebae have as many as 300,000 mitochondria.

And gene loss went much further than a mere 5%. Mitochondria lost nearly all their genes. We have retained just 13 protein-coding genes, along with all other animals. Assuming that the mitochondria derived from ancestors that were not dissimilar to modern α-proteobacteria, they must have started out with around 4,000 genes. Over evolutionary time, they lost more than 99% of their genome. By our calculation above, if 100

---

6 To put some perspective on those numbers, animal cells generally produce actin filaments at a rate of about 1–15 micrometres per *minute*, but some foraminifera can reach speeds of 12 micrometres per *second*. This is the rate of assembly from preformed actin monomers, however, not *de novo* synthesis of actin.

endosymbionts lost 99% of their genes, the energy savings would be close to 1 trillion ATPs over a 24-hour life cycle, or a staggering 12 million per second! But mitochondria don't save energy. They make ATP. Mitochondria are just as good at making ATP as their free-living ancestors, but they reduced the costly bacterial overheads massively. In effect, eukaryotic cells have multibacteria power, but save on the costs of protein synthesis. Or rather, they divert the costs of protein synthesis.

Mitochondria lost most of their genes, but some of them were transferred to the nucleus (more on that in the next chapter). Some of these genes continued to encode the same proteins, carrying out the same old job, so there were no energy savings there. But some of them were no longer needed, either by the host cell or the endosymbiont. They arrived in the nucleus as genetic freebooters, free to change their function, unconstrained as yet by selection. These superfluous stretches of DNA are the genetic raw material for eukaryotic evolution. Some of them spawned whole families of genes that could specialise for new disparate tasks. We know that the earliest eukaryotes had about 3,000 new gene families compared with bacteria. Gene loss from mitochondria enabled the accumulation of new genes in the nucleus at no energetic cost. In principle, if a cell that had 100 endosymbionts transferred 200 genes from each endosymbiont to the nucleus (just 5% of their genes) the host cell would have 20,000 new genes in the nucleus – a whole human genome's worth! – which could be used for all kinds of novel purposes, all at no net energetic cost. The advantage of mitochondria is simply breathtaking.

Two questions remain, and they are tightly linked. First, this entire argument is based on the issue of surface-area-to-volume ratio in prokaryotes. But some bacteria, such as cyanobacteria, are perfectly capable of internalising their bioenergetic membranes, twisting their inner membrane into baroque convolutions, expanding their surface area considerably. Why can't bacteria escape the constraints of chemiosmotic coupling by internalising their respiration in this way? And second, why, if gene loss is so important, did the mitochondria never lose their full genome, taking the process to completion and maximising the energetic benefits of gene loss? The answers to these questions make it clear why bacteria remained stuck in their rut for 4 billion years.

Mitochondria – key to complexity

It's not obvious why mitochondria always retain a handful of genes. Hundreds of genes encoding mitochondrial proteins were transferred to the nucleus early in eukaryotic evolution. Their protein products are now made externally in the cytosol, before being imported into the mitochondria. Yet a small group of genes, encoding respiratory proteins, invariably remained in the mitochondria. Why? The standard textbook *Molecular Biology of the Cell* states: 'We cannot think of compelling reasons why the proteins made in the mitochondria and chloroplasts should be made there, rather than in the cytosol.' That same sentence appears in the 2008, 2002, 1992 and 1983 editions; one is entitled to wonder how much the authors did actually think about the question.

From the standpoint of eukaryotic origins, it seems to me there are two possible types of answer – trivial, or necessary. When I say 'trivial' I don't mean that in a trivial sense – I mean that there is no unmodifiable biophysical reason for the mitochondrial genes to remain where they are. The fact that they have not moved is not because they *can't* move, but because for historical reasons they simply have not. Trivial answers explain why genes *stayed* in the mitochondria: they could have moved to the nucleus, but the balance of chance and selective forces meant that some of them remained where they had always been. Possible reasons include the size and hydrophobicity of mitochondrial proteins, or minor alterations in the genetic code. In principle, the 'trivial' hypothesis argues, all the remaining mitochondrial genes *could* be transferred to the nucleus, albeit requiring a little genetic engineering to modify their sequence as necessary, and the cell would work perfectly well. There are some researchers actively working on transferring mitochondrial genes to the nucleus, on the basis that such a transfer could prevent ageing (more on that in Chapter 7). This is a problem beset with challenges, not a trivial undertaking in the colloquial use of the term; but it is trivial in the sense that these researchers believe there is no need for genes to remain in the mitochondria. They think that there are real benefits to transferring them to the nucleus. Good luck to them.

I disagree with their reasoning. The 'necessary' hypothesis argues that mitochondria have retained genes because they need genes – without them, mitochondria could not exist at all. The cause is unmodifiable: it is not possible to transfer these genes to the nucleus even in principle. Why not? The answer, in my view, comes from John Allen, a biochemist and long-standing colleague. I believe his answer not because he is a friend; quite the reverse. We became friends in part because I believe his answer. Allen has a fertile mind and has put forward a number of original hypotheses, which he has spent decades testing and some of which we have been arguing about for years. In this particular case, he has good evidence supporting the argument that mitochondria (and chloroplasts, for similar reasons) have retained genes because they are needed to control chemiosmotic coupling. Transfer the remaining mitochondrial genes to the nucleus, the argument goes, and the cell will die in time, no matter how carefully crafted the genes may be to their new home. The mitochondrial genes must be right there on site, next to the bioenergetic membranes they serve. I'm told the political term is 'bronze control'.[7] In a war, gold control is the central government, which shapes long-term strategy; silver control is the army command, who plan the distribution of manpower or weaponry used; but a war is won or lost on the ground, under the command of bronze control, the brave men or women who actually engage the enemy, who take the tactical decisions, who inspire their troops, and who are remembered in history as great soldiers. Mitochondrial genes are bronze control, decision-makers on the ground.

Why are such decisions necessary? In Chapter 2 we discussed the sheer power of the proton-motive force. The mitochondrial inner membrane has an electrical potential of about 150–200 millivolts. As the membrane is just 5 nanometres thick, we noted that this translates into a field strength of 30 million volts per metre, equal to a bolt of lightning. Woe betide you if you lose control over such an electrical charge! The penalty is not simply a loss of ATP synthesis, although that alone may well be serious. Failure to transfer electrons properly down the respiratory chains to oxygen (or other

---

7 I was introduced to this term by a former defence secretary, John Reid, who invited me to tea in the House of Lords after reading *Life Ascending*. My attempts to explain the decentralised regulation of mitochondria to my intellectually voracious host turned out to make perfect sense in military terms.

electron acceptors) can result in a kind of electrical short-circuiting, in which electrons escape to react directly with oxygen or nitrogen, to form reactive 'free radicals'. The combination of falling ATP levels, depolarisation of the bioenergetic membranes and release of free radicals is the classic trigger for 'programmed cell death', which we noted earlier is widespread, even in single-celled bacteria. In essence, mitochondrial genes can respond to local changes in conditions, modulating the membrane potential within modest bounds before changes become catastrophic. If these genes were moved to the nucleus, the hypothesis is simply that the mitochondria would lose control over the membrane potential within minutes of any serious changes in oxygen tension or substrate availability, or free-radical leak, and the cell would die.

We need to breathe continuously to stay alive and to exert fine control over muscles in the diaphragm, chest and throat. Down at the level of mitochondria, the mitochondrial genes modulate respiration in much the same way, making sure that output is always finely tailored to demand. No other reason is big enough to explain the universal retention of mitochondrial genes.

This is more than a 'necessary' reason for genes to remain in mitochondria. It is a necessary reason for genes to be stationed next to bioenergetic membranes wherever they may be. It's striking that mitochondria have invariably retained the same small subset of genes in all eukaryotes capable of respiration. On the few occasions that cells lost genes from the mitochondria altogether, they also lost the ability to respire. Hydrogenosomes and mitosomes (the specialised organelles derived from mitochondria found in the archezoa) have generally lost all their genes, and have lost the power of chemiosmotic coupling into the bargain. Conversely the giant bacteria we discussed earlier always have genes (or rather whole genomes) stationed right next to their bioenergetic membranes. For me the case is clinched by cyanobacteria, with their convoluted inner membrane. If genes are necessary to control respiration, then cyanobacteria should have multiple copies of their full genome, in much the same way as the giant bacteria, even though they are substantially smaller. They do. The more complex cyanobacteria often have several hundred copies of their complete genome. As with giant bacteria, that constrains their energy availability per gene – they cannot increase the size of any one genome up to a eukaryotic-sized nuclear genome, because they are obliged instead to accumulate multiple small bacterial genomes.

Here, then, is the reason that bacteria cannot inflate up to eukaryotic size. Simply internalising their bioenergetic membranes and expanding in size does not work. They need to position genes next to their membranes, and the reality, in the absence of endosymbiosis, is that those genes come in the form of full genomes. There is no benefit in terms of energy per gene from becoming larger, except when large size is attained by endosymbiosis. Only then is gene loss possible, and only then can the shrinking of mitochondrial genomes fuel the expansion of the nuclear genome over several orders of magnitude, up to eukaryotic sizes.

You might have thought of another possibility: the use of bacterial plasmids, semi-independent rings of DNA that can carry scores of genes on occasions. Why couldn't the genes for respiration be placed on one large plasmid, and then multiple copies of this plasmid be stationed next to the membranes? There are potentially intractable logistical difficulties with this, but could it work in principle? I think not. Among prokaryotes there is no advantage to being bigger for its own sake, and no advantage to having more ATP than necessary. Small bacteria are not short of ATP: they have plenty. Being a little larger and having a little more ATP carries no benefit; it is better to be a little smaller and have just enough ATP – and replicate faster. A second disadvantage to expanding in volume for its own sake is that supply lines are needed to serve remote regions of the cell. A large cell needs to ship cargo to all quarters, and eukaryotes do exactly that. But such transport systems do not evolve overnight. That takes many generations, during which time there would need to be some other advantage to being bigger. So plasmids won't work – they put the cart before the horses. By far the simplest solution to the problem of distribution is just to side-step it altogether, to have multiple copies of a full genome, each controlling a 'bacterial' volume of cytoplasm, as in the giant bacteria.

So how did eukaryotes break out of the size loop, and evolve complex transport systems? What is so different about a large cell with multiple mitochondria, each one of which has its own plasmid-sized genome, and a giant bacterium with multiple plasmids, dispersed to control respiration?

The answer is that the deal at the origin of eukaryotes had nothing to do with ATP, as pointed out by Bill Martin and Miklos Müller in their hypothesis for the first eukaryote. Martin and Müller propose a metabolic syntrophy between the host cell and its endosymbionts, meaning that they trade in the substrates of growth, not just energy. The hydrogen hypothesis argues that the first endosymbionts provided their methanogen host cells with hydrogen needed for growth. We don't need to worry about the details here. The point is that without their substrate (hydrogen in this case) the host cells cannot grow at all. The endosymbionts provide *all* the substrate needed for growth. The more endosymbionts, the more substrate, the faster the host cells can grow; and the better the endosymbionts do too. In the case of endosymbioses, then, larger cells do benefit because they contain more endosymbionts, and so gain more fuel for growth. They will do even better as they develop transport networks to their own endosymbionts. This almost literally puts the horses (power supply) before the cart (transport).

As the endosymbionts lose genes, their own ATP demands fall. There's an irony here. Cell respiration produces ATP from ADP, and as the ATP is broken back down to ADP it powers work around the cell. If ATP is not consumed, then the entire pool of ADP is converted into ATP, and respiration grinds to a halt. Under these conditions the respiratory chain accumulates electrons, becoming highly 'reduced' (more on this in Chapter 7). It is then reactive with oxygen, leaking free radicals that can damage the surrounding proteins and DNA, or even trigger cell death. The evolution of one key protein, the ADP–ATP transporter, enabled the host cell to bleed off the endosymbionts' ATP for its own purposes, but tellingly, also solved this problem for the endosymbionts. By bleeding off excess ATP and resupplying the endosymbionts with ADP, the host cell restricted free-radical leak within the endosymbiont, and so lowered the risk of damage and cell death. This helps explain why it was in the interests of both the host cell and endosymbionts to 'burn' ATP on extravagant building projects such as a dynamic cytoskeleton.[8] But the key point is that there were advantages at

---

8 There is an instructive bacterial precedent for burning ATP, known as ATP or energy 'spilling'. The term is accurate: some bacteria can splash away up to two-thirds of their

every stage of the endosymbiotic relationship, unlike plasmids, which offer no benefit to being larger or having more ATP for its own sake.

The origin of the eukaryotic cell was a singular event. Here on earth it happened just once in 4 billion years of evolution. When considered in terms of genomes and information, this peculiar trajectory is nearly impossible to understand. But when considered in terms of energy and the physical structure of cells it makes a great deal of sense. We have seen how chemiosmotic coupling may have arisen in alkaline hydrothermal vents, and why it remained universal in bacteria and archaea for all eternity. We have seen that chemiosmotic coupling made possible the wonderful adaptability and versatility of prokaryotes. Such factors are likely to play out on other planets too, right back to the beginnings of life from little more than rock, water and CO_2. Now we see, too, why natural selection, operating on infinite populations of bacteria over infinite periods of time, should not give rise to large complex cells, what we know as eukaryotes, except by way of a rare and stochastic endosymbiosis.

There is no innate or universal trajectory towards complex life. The universe is not pregnant with the idea of ourselves. Complex life might arise elsewhere, but it is unlikely to be common, for the same reasons it did not arise repeatedly here. The first part of the explanation is simple – endosymbioses between prokaryotes are not common (although we do know of a couple of examples, so we know that they can happen). The second part is less obvious, and smacks of Sartre's vision of hell as other people. The intimacy of endosymbiosis might have broken the endless deadlock of bacteria, but in the next chapter we shall see that the tormented birth of this new entity, the eukaryotic cell, goes some way towards explaining why such events happen very rarely, and why all complex life shares so many peculiar traits, from sex to death.

---

overall ATP budget on futile cycling of ions across the cell membrane and other equally pointless feats. Why? One possible answer is that it keeps a healthy balance of ATP to ADP, which keeps the membrane potential and free-radical leak under control. Again, it goes to show that bacteria have plenty of ATP to spare – they are not in any way energetically challenged; only scaling up to eukaryotic sizes reveals the energy-per-gene problem.

6

SEX AND THE ORIGINS OF DEATH

Nature abhors a vacuum, said Aristotle. The idea was echoed, two millennia later, by Newton. Both worried about what fills space; Newton believed it was a mysterious substance known as the æther. In physics, the idea fell into disrepute in the twentieth century, but the *horror vacui* retains all its strength in ecology. The filling up of every ecological space is nicely captured in an old rhyme: 'Big fleas have little fleas upon their backs to bite 'em; little fleas have smaller fleas, and so ad infinitum.' Every conceivable niche is occupied, with each species exquisitely adapted to its own space. Every plant, every animal, every bacterium, is a habitat in itself, a jungle of opportunities for all kinds of jumping genes, viruses and parasites, to say nothing of big predators. Anything and everything goes.

Except that it doesn't. It only looks that way. The infinite tapestry of life is but a semblance, with a black hole at its heart. It is time to address the greatest paradox in biology: why it is that all life on earth is divided into prokaryotes, which lack morphological complexity, and eukaryotes, which share a massive number of detailed properties, none of which are found in prokaryotes. There is a gulf, a void, a vacuum, between the two, which nature really ought to abhor. All eukaryotes share more or less everything; all prokaryotes have, from a morphological point of view, next to nothing. There is no better illustration of the inequitable biblical tenet 'to him that hath shall be given'.

In the previous chapter, we saw that an endosymbiosis between two prokaryotes broke the endless loop of simplicity. It is not easy for one bacterium to get inside another one and to survive there for endless generations, but we know of a few examples, so we know that it does happen, if very rarely. But a cell within a cell was just the beginning, a pregnant moment in the history of life, yet no more than that. It is just a cell within a cell. Somehow we have to chart a course from there to the birth of true complexity – to a cell that has accumulated everything common to all eukaryotes. We start with bacteria, lacking almost all complex traits, and end with complete eukaryotes, cells with a nucleus, a plethora of internal membranes and compartments, a dynamic cell skeleton and complex behaviour such as sex. Eukaryotic cells expanded in genome size and in physical size over four or five orders of magnitude. The last common ancestor of eukaryotes had accumulated all these traits; the starting point, a cell within a cell, had none of them. There are no surviving intermediates, nothing much to tell us how or why any of these complex eukaryotic traits evolved.

It's sometimes said that the endosymbiosis that launched the eukaryotes was not Darwinian: that it was not a gradual succession of small steps but a sudden leap into the unknown, creating a 'hopeful monster'. To a point that is true. I have argued that natural selection, acting on infinite populations of prokaryotes over infinite periods of time, will never produce complex eukaryotic cells except by way of an endosymbiosis. Such events cannot be represented on a standard bifurcating tree of life. Endosymbiosis is bifurcation backwards, where the branches do not branch but fuse together. But an endosymbiosis is a singular event, a moment in evolution that can't produce a nucleus or any of the other archetypal eukaryotic traits. What it did do was set in motion a train of events, which are perfectly Darwinian in the normal sense of the word.

So I am not arguing that the origin of eukaryotes was non-Darwinian but that the selective landscape was transformed by a singular endosymbiosis between prokaryotes. After that it was Darwin all the way. The question is, how did the acquisition of endosymbionts alter the course of natural selection? Did it happen in a predictable manner, which might follow a similar course on other planets, or did the elimination of energetic constraints open the floodgates to unfettered evolution? I shall argue that at least some of the universal traits of eukaryotes were wrought in the intimate relationship between host cell and endosymbiont, and as such are predictable from first principles. These traits include the nucleus, sex, two sexes, and even the immortal germline, begetter of the mortal body.

Starting out with an endosymbiosis immediately places some constraints on the order of events; the nucleus and membrane systems must have arisen after the endosymbiosis, for example. But it also places some constraints on the speed at which evolution must have operated. Darwinian evolution and gradualism are easily conflated, but what does 'gradual' actually mean? It means simply that there are no great leaps into the unknown, that all *adaptive* changes are small and discrete. That is not true if we consider changes to the genome itself, which might take the form of large deletions, duplications, transpositions or abrupt rewiring as a result of regulatory genes being inappropriately switched on or off. But such changes are not adaptive; like endosymbioses, they merely alter the starting point from which selection acts. To suggest that the nucleus, for example, somehow just popped into existence is to confound genetic saltation with adaptation. The nucleus is an exquisitely adapted structure, no mere repository for DNA. It is composed of structures such as the nucleolus, where new ribosomal RNA is manufactured on a colossal scale; the doubled nuclear membrane, studded with stunningly beautiful protein pore complexes (**Figure 26**), each one containing scores of proteins conserved across all eukaryotes; and the elastic lamina, a flexible protein meshwork lining the nuclear membrane that protects DNA against shear stress.

The point is that such a structure is the product of natural selection acting over extended periods of time, and requires the refinement and orchestration of hundreds of separate proteins. All of this is a purely Darwinian process. But that does not mean it had to happen slowly in geological terms. In the fossil record, we are used to seeing long periods of stasis, punctuated occasionally by periods of fast change. This change is fast in geological time, but not necessarily in terms of generations: it is simply not hampered by the same constraints that oppose change under normal circumstances. Only rarely is natural selection a force for change. Most commonly, it

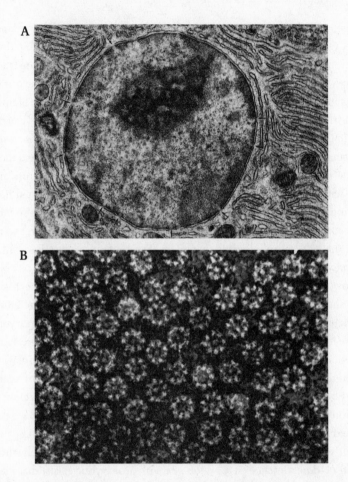

Figure 26 **Nuclear pores**
Classic images by the pioneer of electron microscopy Don Fawcett. The double membrane surrounding the eukaryotic nucleus is clearly visible, as are the regular pores, marked by arrows in **A**. The darker areas within the nucleus are relatively inactive regions, where chromatin is 'condensed', whereas lighter regions indicate active transcription. The lighter 'spaces' close to the nuclear pores indicate active transport in and out of the nucleus. **B** shows an array of nuclear pore complexes, each composed of scores of proteins assembled to form the machinery of import and export. The core proteins in these pore complexes are conserved across all eukaryotes, hence nuclear pores must have been present in LECA (the last eukaryotic common ancestor). [Don Fawcett/Photo Researchers]

opposes change, purging variations from the peaks of an adaptive land-scape. Only when that landscape undergoes some kind of seismic shift does selection promote change rather than stasis. And then it can operate start-lingly swiftly. The eye is a good example. Eyes arose in the Cambrian explo-sion, apparently within the space of a couple of million years. When blunted to the rhythm of hundreds of millions of years during the near-eternal Precambrian, 2 million years seems indecently hasty. Why stasis for so long, then such rapid-fire change? Perhaps because oxygen levels rose, and then, for the first time, selection favoured large active animals, predators and prey, with eyes and shells.[1] A famous mathematical model calculated how long it might take for an eye to evolve from a simple light-sensitive spot on some sort of worm. The answer, assuming a life cycle of one year, and no more than 1% morphological change in each generation, was just half a million years.

How long should it take a nucleus to evolve? Or sex, or phagocytosis? Why should it take any longer than the eye? This is a project for the future – to calculate the minimum time to evolve a eukaryote from a prokaryote. Before it's worth embarking on such a project, we need to know more about the sequence of events involved. But there is no *prima facie* reason to assume it should take vast tracts of time measured in hundreds of millions of years. Why not 2 million years? Assuming one cell division per day, that's close to a billion generations. How many are needed? Once the energetic brakes that blocked the evolution of complexity in prokaryotes were lifted, I see no reason why eukaryotic cells could not have evolved in a relatively short period of time. Set against 3 billion years of prokaryotic stasis, that mas-querades as a sudden leap forward; but the process was strictly Darwinian.

Just because it is conceivable for evolution to operate quickly does not mean that it actually did. But there are strong grounds to think that the

---

1 I'm not claiming that a rise in oxygen concentration drove the evolution of animals (as discussed in Chapter 1), but that it enabled more active behaviour in large animals. The release from energetic constraints promoted a polyphyletic radiation of many different groups of animals, but animals had already evolved before the Cambrian explosion, before the major rise in oxygen towards the end of the Precambrian.

evolution of the eukaryotes probably did happen quickly, based on nature's abhorrence of a vacuum. The problem is precisely the fact that eukaryotes share everything, and prokaryotes have none of it. That implies instability. In Chapter 1, we considered the archezoa, those relatively simple single-celled eukaryotes that were once mistaken for evolutionary intermediates between prokaryotes and eukaryotes. This disparate group turned out to be derived from more complex ancestors with a full stock of all eukaryotic traits. But they are nonetheless true *ecological* intermediates – they occupy the niche of morphological complexity between prokaryotes and eukaryotes. They fill the vacuum. To a superficial first glance, then, there is no vacuum: there is a continuous spectrum of morphological complexity ranging from parasitic genetic elements to giant viruses, bacteria to simple eukaryotes, complex cells to multicellular organisms. Only recently, when it transpired that the archezoa are a sham, did the full horror of the vacuum become evident.

The fact that the archezoa were not outcompeted to extinction means that simple intermediates can thrive in this space. There is no reason why the same ecological niche could not have been occupied by genuine evolutionary intermediates, cells that lacked mitochondria, or a nucleus, or peroxisomes, or membrane systems such as the Golgi apparatus or endoplasmic reticulum. If the eukaryotes arose slowly, over tens or hundreds of millions of years, there must have been many stable intermediates, cells that lacked various eukaryotic traits. They should have occupied the same intermediate niches now filled with archezoa. Some of them ought to have survived until today, as genuine evolutionary intermediates in the vacuum. But no! None are to be found, despite a long, hard look. If they were not outcompeted to extinction, then why did none of them survive? I would say because they were genetically unstable. There were not many ways to cross the void, and most perished.

That would imply a small population size, which also makes sense. A large population indicates evolutionary success. If the early eukaryotes were thriving, they should have spread out, occupied new ecological spaces, diverged. They should have been genetically stable. At least some of them should have survived. But that didn't happen. At face value, then, it seems most likely that the first eukaryotes were genetically unstable, and evolved quickly in a small population.

There's another reason to think that this must be true: the fact that all eukaryotes share exactly the same traits. Think about how peculiar this is! We all share the same traits with other human beings, such as upright posture, furless bodies, opposing thumbs, large brains and a facility for language, as we are all related by ancestry and interbreeding. Sex. That is the simplest definition of a species – a population of interbreeding individuals. Groups that do not interbreed diverge, and evolve distinct traits – they become new species. Yet this didn't happen at the origin of eukaryotes. All eukaryotes share the same set of basic traits. It looks a lot like an interbreeding population. Sex.

Could any other form of reproduction have achieved the same end point? I don't think so. Asexual reproduction – cloning – leads to deep divergence, as different mutations accumulate in different populations. These mutations are subject to selection in disparate environments, facing different advantages and disadvantages. Cloning may produce identical copies, but ironically this ultimately drives divergence between populations as mutations accumulate. In contrast, sex pools traits in a population, forever mixing and matching, opposing divergence. The fact that eukaryotes share the same traits suggests that they arose in an interbreeding sexual population. This in turn implies that their population was small enough to interbreed. Any cells that did not have sex, in this population, did not survive. The Bible was right: 'Strait is the gate and narrow is the way, which leadeth unto life, and few there be that find it.'

What about lateral gene transfer, rife as it is in bacteria and archaea? Like sex, lateral gene transfer involves recombination, producing 'fluid' chromosomes with shifting combinations of genes. Unlike sex, though, lateral gene transfer is not reciprocal, and does not involve cell fusion or recombination across the full genome. It is piecemeal and unidirectional: it does not combine traits in a population, but increases divergence between individuals. Just consider *E. coli*. A single cell may contain about 4,000 genes, but the 'metagenome' (the total number of genes found in different strains of *E. coli*, as defined by ribosomal RNA) is more like 18,000 genes. The outcome of rampant lateral gene transfer is that different strains differ in up to half of their genes – more variation than in all the vertebrates put together. In short, neither cloning nor lateral gene transfer, the dominant modes of inheritance in bacteria and archaea, can explain the enigma of uniformity in eukaryotes.

If I were writing this a decade ago, the idea that sex arose very early in eukaryotic evolution would have had little evidence supporting it; numerous species, including many amoebae and supposedly deep-branching archezoa such as *Giardia*, were taken to be asexual. Even now, nobody has caught *Giardia* in flagrante, in the act of microbial sex. But what we lack in natural history, we make up for in technology. We know its genome sequence. It contains the genes needed for meiosis (reductive cell division to produce gametes for sex) in perfect working order, and the structure of its genome bears witness to regular sexual recombination. The same goes for more or less every other species we have looked at. With the exception of secondarily derived asexual eukaryotes, which usually fall extinct quickly, all known eukaryotes are sexual. We can take it that their common ancestor was too. In sum: sex arose very early in eukaryotic evolution, and *only* the evolution of sex in a small unstable population can explain why all eukaryotes share so many common traits.

That brings us to the question of this chapter. Is there something about an endosymbiosis between two prokaryotes that might drive the evolution of sex? You bet, and much else besides.

The secret in the structure of our genes

Eukaryotes have 'genes in pieces'. Few discoveries in twentieth-century biology came as a greater surprise. We had been misled by early studies on bacterial genes to think that genes are like beads on a string, all lined up in a sensible order on our chromosomes. As the geneticist David Penny put it: 'I would be quite proud to have served on the committee that designed the *E. coli* genome. There is, however, no way that I would admit to serving on the committee that designed the human genome. Not even a university committee could botch something that badly.'

So what went wrong? Eukaryotic genes are a mess. They are composed of relatively short sequences that code for bits of proteins, broken up by long tracts of non-coding DNA, known as introns. There are typically several introns per gene (which is usually defined as a stretch of DNA encoding a single protein). These vary enormously in length, but are often substantially longer than the protein-coding sequences themselves. They are always copied into the RNA template that specifies the sequence of amino acids in the protein, but are then spliced out before the RNA reaches the ribosomes, the great protein-building factories in the cytoplasm. This is no easy task. It is achieved by another remarkable protein nanomachine known as the spliceosome. We'll return to the significance of the spliceosome soon. For now, let's just note that the whole procedure is a weirdly roundabout way of going about things. Any failure to splice out these introns means that reams of nonsensical RNA code is fed into the ribosomes, which go right ahead and synthesise the nonsensical proteins. The ribosomes are as beholden to their red tape as a Kafka bureaucrat.

Why do eukaryotes have genes in pieces? There are a few known benefits. Different proteins can be pieced together from the same gene by differential splicing, enabling the recombinatorial virtuosity of the immune system, for example. Different bits of protein are recombined in marvellous ways to form billions of distinct antibodies, which are capable of binding to practically any bacterial or viral protein, thereby setting in motion the killing machines of the immune system. But immune systems are late inventions of large, complex animals. Was there an earlier advantage? In the 1970s, one of the doyens of twentieth-century evolutionary biology, Ford Doolittle, suggested that introns might date back to the very origins of life on earth – an idea known as the 'introns early' hypothesis. The idea was that early genes, lacking sophisticated modern DNA repair machinery, must have accumulated errors very rapidly, making them extremely prone to mutational meltdown. Given a high mutation rate, the number of mutations that accumulate depends on the length of DNA. Only small genomes could possibly avoid meltdown. Introns were an answer. How to encode a large number of proteins with a short stretch of DNA? Just recombine small bits and pieces. It's a beautiful notion, which still retains a few adherents, if not

Doolittle himself. The hypothesis, like all good hypotheses, makes a number of predictions; unfortunately, these turn out not to be true.

The major prediction is that eukaryotes must have evolved first. Only eukaryotes have true introns. If introns were the ancestral state, then eukaryotes must have been the earliest cells, preceding the bacteria and archaea, which must have lost their introns later on by selection for streamlining their genomes. That makes no phylogenetic sense. The modern era of whole-genome sequencing shows incontrovertibly that eukaryotes arose from an archaeal host cell and a bacterial endosymbiont. The deepest branch in the tree of life is between archaea and bacteria; eukaryotes arose more recently, a view that is also consistent with the fossil record and the energetic considerations of the last chapter.

But if introns are not an ancestral state, where did they come from, and why? The answer seems to be the endosymbiont. I said that 'true introns' are not found in bacteria, but their precursors almost certainly are bacterial, or rather, bacterial genetic parasites, technically termed 'mobile group II self-splicing introns'. Don't worry about words. Mobile introns are just bits of selfish DNA, jumping genes which copy themselves in and out of the genome. But I shouldn't say 'just'. They are remarkable and purposeful machines. They are read off into RNA in the normal fashion, but then spring to life (what other word is there?), forming themselves into pairs of RNA 'scissors'. These splice out the parasites from the longer RNA transcripts, minimising damage to the host cell, to form active complexes that encode a reverse transcriptase – an enzyme capable of converting RNA back into DNA. These insert copies of the intron back into the genome. So introns are parasitic genes, which splice themselves in and out of bacterial genomes.

'Big fleas have little fleas upon their backs to bite 'em…' Who would have thought that the genome is a snake pit, seething with ingenious parasites that come and go at their pleasure. But that's what it is. These mobile introns are probably ancient. They are found in all three domains of life, and unlike viruses they never need to leave the safety of their host cell. They are copied faithfully each time the host cell divides. Life has just learnt to live with them.

And bacteria are quite capable of dealing with them. We don't quite know how. It might simply be the strength of selection acting on large populations. Bacteria with badly positioned introns, which interfere with their genes in some way, simply lose out in the selective battle with cells that do not have badly positioned introns. Or perhaps the introns themselves are accommodating, and invade peripheral regions of DNA that don't upset their host cells much. Unlike viruses, which can survive on their own, and so don't care much about killing their host cells, mobile introns perish with their hosts, so gain nothing from obstructing them. The language that lends itself best to analysing this kind of biology is that of economics: the mathematics of costs and benefits, the prisoner's dilemma, game theory. Be that as it may, the fact is that mobile introns do not run rife in bacteria or archaea, and are not found within the genes themselves – they are therefore not technically introns at all – but accumulate at low density in intergenic regions. A typical bacterial genome is unlikely to contain more than about 30 mobile introns (in 4,000 genes) compared with tens of thousands of introns in eukaryotes. The low number of introns in bacteria reflects the long-term balance of costs and benefits, the outcome of selection acting on both parties over many generations.

This is the kind of bacterium that entered into an endosymbiosis with an archaeal host cell 1.5 to 2 billion years ago. The closest modern equivalent is an α-proteobacterium of some sort, and we know that modern α-proteobacteria contain low numbers of mobile introns. But what connects these ancient genetic parasites with the structure of eukaryotic genes? Little more than the detailed mechanism of the RNA scissors that splice out mobile bacterial introns, and simple logic. I mentioned the spliceosomes a few paragraphs ago: these are the protein nanomachines that cut out the introns from our own RNA transcripts. The spliceosome is not only made of proteins: at its heart is a pair of RNA scissors, the very same. These splice out eukaryotic introns by way of a telltale mechanism that betrays their ancestry as bacterial self-splicing introns (**Figure 27**).

That's it. There is nothing about the genetic sequence of the introns themselves to suggest that they derive from bacteria. They do not encode proteins such as reverse transcriptase, they do not splice themselves in and

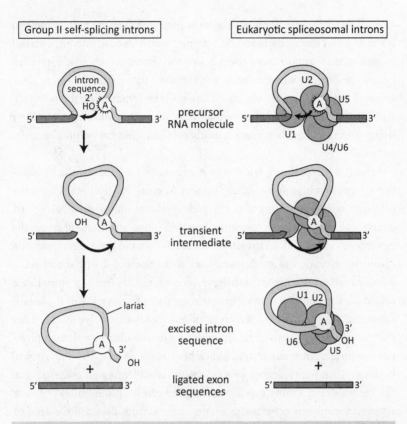

| Group II self-splicing introns | Eukaryotic spliceosomal introns |

Figure 27 **Mobile self-splicing introns and the spliceosome**
Eukaryotic genes are composed of exons (sequences that encode proteins) and introns – long, non-coding sequences inserted into genes, which are spliced out from the RNA code-script before the protein is synthesised. Introns seem to be derived from parasitic DNA elements found in bacterial genomes (left panel), but decayed by mutations to inert sequences in eukaryotic genomes. These must be actively removed by the spliceosome (right panel). The rationale for this argument is the mechanism of splicing shown here. The bacterial parasite (left panel) splices itself out to form an excised intron sequence encoding a reverse transcriptase that can convert copies of the parasitic genes into DNA sequences, and insert multiple copies into the bacterial genome. The eukaryotic spliceosome (right panel) is a large protein complex, but its function depends on a catalytic RNA (ribozyme) at its heart, which shares exactly the same mechanism of splicing. This suggests that the spliceosome, and by extension eukaryotic introns, derived from mobile group II self-splicing introns released from the bacterial endosymbiont early in eukaryotic evolution.

out of DNA, they are not mobile genetic parasites, they are merely lumpen tracts of DNA that sit there and do nothing.[2] But these dead introns, decayed by mutations that punctured them below the waterline, are now corrupted beyond all recognition, and far more dangerous than living parasites. They can no longer cut themselves out. They must be removed by the host cell. And so they are, using scissors that were once requisitioned from their living cousins. The spliceosome is a eukaryotic machine based on a bacterial parasite.

Here is the hypothesis, laid out in an exciting 2006 paper by the Russian-born American bioinformatician, Eugene Koonin, and Bill Martin. At the origin of eukaryotes, they said, the endosymbiont unleashed a barrage of genetic parasites upon the unwitting host cell. These proliferated across the genome in an early intron invasion, which sculpted eukaryotic genomes and drove the evolution of deep traits such as the nucleus. I would add sex. I admit all this sounds like make-believe, an evolutionary 'just-so' story based on the flimsy evidence of an incriminating pair of scissors. But the idea is supported by the detailed structure of the genes themselves. The sheer number of introns – tens of thousands of them – combined with their physical position within eukaryotic genes bear mute witness to their ancient heritage. That heritage goes beyond the introns themselves, and speaks to the tortured and intimate relationship between host and endosymbiont. Even if these ideas are not the whole truth, I think they are the *kind* of answers we seek.

---

2 OK, mostly nothing. Some introns have acquired functions, such as binding transcription factors, and sometimes they are active as RNA themselves, interfering with protein synthesis and the transcription of other genes. We are in the midst of an era-defining argument about the function of non-coding DNA. Some of it is certainly functional, but I align myself with the doubters, who argue that most of the (human) genome is not actively constrained in its sequences, and therefore does not serve a purpose that is defined by its sequence. To all intents and purposes, that means it doesn't have a function. If forced to hazard a guess, I would say that perhaps 20% of the human genome is functional, and the rest is basically junk. But that doesn't mean it is not useful for some other purpose, such as filling space. Nature abhors a vacuum, after all.

Introns and the origin of the nucleus

The positions of many introns are conserved across eukaryotes. This is another unexpected curiosity. Take a gene encoding a protein that is involved in basic cell metabolism found in all eukaryotes, for example citrate synthase. We'll find the same gene in ourselves, as well as in seaweeds, mushrooms, trees and amoebae. Despite diverging somewhat in sequence over the incomprehensible number of generations that separate us from our common ancestor with trees, natural selection has acted to conserve its function, and thus its specific gene sequence. This is a beautiful illustration of shared ancestry, and the molecular basis of natural selection. What nobody expected is that such genes should typically contain two or three introns, frequently inserted in exactly the same positions in trees and humans. Why should that be? There are only two plausible explanations. Either the introns inserted themselves in the same places independently, because those particular sites were favoured by selection for some reason, or they inserted themselves once into the common ancestor of eukaryotes and were then passed down to their descendants. Some of these descendants may have lost them again, of course.

If there were only a handful of cases known, we might favour the former interpretation, but the fact that thousands of introns are inserted into exactly the same positions in hundreds of shared genes across all eukaryotes makes this seem implausible. Shared ancestry is much the most parsimonious explanation. If so, then there must have been an early wave of intron invasion, soon after the origin of the eukaryotic cell, which was responsible for implanting all these introns in the first place. Then after that there must have been some kind of mutational corruption of introns, which robbed them of their mobility, preserving their positions in all later eukaryotes like the indelible chalk outlines of corpses.

There is also another more compelling reason to favour an early intron invasion. We can distinguish between different types of gene, known as *orthologs* and *paralogs*. Orthologs are basically the same genes doing the same job in different species, inherited from a common ancestor, as in the example we've just considered. So all eukaryotes have an ortholog of the gene for citrate synthase, which we all inherited from our common ancestor.

The second group of genes, paralogs, also share a common ancestor, but in this case the ancestral gene was duplicated *within the same cell*, often on multiple occasions, to give a gene family. Such families can contain as many as 20 or 30 genes, each of which usually ends up specialising to a slightly different task. An example is the haemoglobin family of about 10 genes, all of which encode very similar proteins, with each serving a slightly different purpose. In essence, orthologs are equivalent genes in different species, while paralogs are members of a gene family in the same organism. But of course entire families of paralogs can also be found in different species, inherited from their common ancestor. So all mammals have paralogous haemoglobin gene families.

We can break down these families of paralogous genes into either ancient or recent paralogs. In an ingenious study, Eugene Koonin did exactly this. He defined *ancient* paralogs as gene families that are found in all eukaryotes, but which are not duplicated in any prokaryotes. We can therefore place the round of gene duplications that gave rise to the gene family as an early event in eukaryotic evolution, before the evolution of the last eukaryotic common ancestor. *Recent* paralogs, in contrast, are gene families found only in certain eukaryotic groups, such as animals or plants. In this case, we can conclude that the duplications occurred more recently, during the evolution of that particular group.

Koonin predicted that if there was indeed an intron invasion during early eukaryotic evolution, then mobile introns should have randomly inserted themselves into different genes. That's because the ancient paralogs were being actively duplicated during this same period. If the early intron invasion had not yet abated, then mobile introns would still be inserting themselves into new positions in different members of the growing paralogous gene family. In contrast, more recent duplications of paralogs occurred well after the end of the postulated early intron invasion. With no new insertions, the old intron positions should be conserved in new copies of these genes. In other words, ancient paralogs should have poor conservation of intron position relative to recent paralogs. This is true to a remarkable degree. Practically all intron positions are conserved in recent paralogs, whereas there is very poor intron conservation in ancient paralogs, exactly as predicted.

All this suggests that early eukaryotes really did suffer an invasion of mobile introns from their own endosymbionts. But if so, why did these proliferate in early eukaryotes when they are normally kept under tight control in both bacteria and archaea? There are two possible answers, and the chances are that both are true. The first reason is that early eukaryotes – basically still prokaryotes, *archaea* – suffered a bombardment of *bacterial* introns from uncomfortably close quarters, from inside their own cytoplasm. There is a ratchet operating here. An endosymbiosis is a natural 'experiment' which might fail. If the host cell dies, the experiment is over. But that's not true the other way around. If there is more than one endosymbiont, and just one of them dies, the experiment continues – the host cell survives, with all its other endosymbionts. But the dead endosymbiont's DNA spills into the cytosol, whence it is likely to be recombined into the host cell's genome by standard lateral gene transfer.

This is not easy to stop, and continues to this day – our nuclear genomes are riddled with thousands of bits and pieces of mitochondrial DNA, called '*numts*' (nuclear mitochondrial sequences, since you ask), which arrived there by exactly such transfers. New *numts* crop up occasionally, calling attention to themselves when they disrupt a gene, causing a genetic disease. Back at the origin of eukaryotes, before there was a nucleus at all, such transfers must have been more common. The chaotic transfer of DNA from mitochondria to host cell would have been worse if selective mechanisms *do* exist that direct mobile introns to particular sites within a genome, while avoiding others. In general, bacterial introns are adapted to their bacterial hosts, and archaeal introns to their archaeal hosts. In the early eukaryotes, however, *bacterial* introns were invading an *archaeal* genome, with very different gene sequences. There were no adaptive constraints; and without them, what could have stopped introns from proliferating uncontrollably? Nothing! Extinction loomed. The best we could hope for is a small population of genetically unstable – sickly – cells.

The second reason for an early intron proliferation is the low strength of selection acting against it. In part, this is precisely because a small population of sickly cells is less competitive than a heaving population of healthy cells. But the first eukaryotes should also have had an unprecedented tolerance for intron invasion. After all, their source was the endosymbiont, the future mitochondria, which are an energetic boon as well as a genetic cost. Introns are a cost to bacteria because they are an energetic and genetic burden; small cells with less DNA replicate faster than large cells with more DNA than they need. As we saw in the last chapter, bacteria streamline their genomes to a minimum compatible with survival. In contrast, eukaryotes exhibit extreme genomic asymmetry: they are free to expand their nuclear genomes precisely because their endosymbiont genomes shrink. Nothing is planned about the expansion of the host cell genome; it is simply that increased genome size is not penalised by selection in the same way as it is in bacteria. This limited penalty enables eukaryotes to accumulate thousands more genes, through all kinds of duplications and recombinations, but also to tolerate a far heavier load of genetic parasites. The two must inevitably go hand in hand. Eukaryotic genomes became overrun with introns because, from an energetic point of view, they could be.

So it seems likely that the first eukaryotes suffered a bombardment of genetic parasites from their own endosymbionts. Ironically, these parasites didn't pose much of a problem: the problem really began when the parasites decayed and died, leaving their corpses – introns – littering the genome. Now the host cell had to physically cut the introns out, or they would be read off into nonsensical proteins. As we've noted, this is done by the spliceosome, which derives from the RNA scissors of mobile introns. But the spliceosome, impressive nanomachine though it might be, is only a partial solution. The trouble is that spliceosomes are slow. Even today, after nearly 2 billion years of evolutionary refinement, they take several minutes to cut out a single intron. In contrast, ribosomes work at a furious pace – up to 10 amino acids per second. It takes barely half a minute to make a standard bacterial protein, about 250 amino acids in length. Even if the spliceosome could gain access to RNA (which is not easy as RNA is often encrusted in multiple ribosomes) it could not stop the formation of a large number of useless proteins, with their introns incorporated intact.

How could an error catastrophe be averted? Simply by inserting a barrier in the way, according to Martin and Koonin. The nuclear membrane is a barrier separating transcription from translation – inside the nucleus, genes are transcribed into RNA codescripts; outside the nucleus, the RNAs are translated into proteins on the ribosomes. Crucially, the slow process of splicing takes place inside the nucleus, before the ribosomes can get anywhere near the RNA. That is the whole point of the nucleus: to keep ribosomes at bay. This explains why eukaryotes need a nucleus but prokaryotes don't – prokaryotes don't have an intron problem.

But hang on a minute, I hear you cry! We can't just pull out a perfectly formed nuclear membrane from nowhere! It must have taken many generations to evolve, so why didn't the early eukaryotes die out in the meantime? Well, no doubt many did, but the problem might not be as difficult as all that. The key lies in another curiosity relating to membranes. Even though it is clear from the genes that the host cell was a *bona fide* archaeon, which must have had characteristic *archaeal* lipids in its membranes, eukaryotes have *bacterial* lipids in their membranes. That's a fact to conjure with. For some reason, the archaeal membranes must have been replaced with bacterial membranes early on in eukaryotic evolution. Why?

There are two facets to this question. The first is a matter of practicality: could it actually be done? The answer is yes. Rather surprisingly, mosaic membranes, composed of various different mixtures of archaeal and bacterial lipids, are in fact stable; we know this from lab experiments. It is therefore possible to transit gradually from an archaeal to a bacterial membrane. So there's no reason why it shouldn't happen, but the fact is that such transitions are rare indeed. That brings us to the second facet – what rare evolutionary force might drive such a change? The answer is the endosymbiont.

The chaotic transfer of DNA from endosymbionts to the host cell must have included the genes for bacterial lipid synthesis. We can take it that the enzymes encoded were synthesised and active: they went right ahead and made bacterial lipids, but initially this synthesis was likely to be uncontrolled. What happens when lipids are synthesised in a random fashion? When formed in water, they simply precipitate as lipid vesicles. Jeff Errington in Newcastle has shown that real cells behave in the same way: mutations that increase lipid synthesis in bacteria result in precipitation of internal membranes. These tend to precipitate close to where they are formed, surrounding the genome with piles of lipid 'bags'. Just as a tramp might insulate himself from the cold with plastic bags, if inadequately, so piles of lipid bags would ease the intron problem by providing an imperfect barrier between the DNA and ribosomes. That barrier *needed* to be imperfect. A sealed membrane would prevent export of RNA out to the ribosomes. A broken barrier would merely slow it down, giving the spliceosomes a little more time to cut out the introns before the ribosomes could get to work. In other words, a random (but predictable) starting point gave selection the beginnings of a solution. The beginning was a pile of lipid bags surrounding a genome; the end point was the nuclear membrane, replete with its sophisticated pores.

The morphology of the nuclear membrane is consistent with this view. Lipid bags, like plastic bags, can be flattened. In cross section, a flattened bag has two closely aligned parallel sides – a double membrane. That is precisely the structure of the nuclear membrane: a series of flattened vesicles, fused together, with the nuclear pore complexes nestling in the interstices. During cell division the membrane disintegrates back into separate small vesicles; afterwards, they grow and fuse again to reconstitute the nuclear membranes of the two daughter cells.

The pattern of genes encoding nuclear structures also makes sense in this light. If the nucleus had evolved *before* the acquisition of mitochondria, then the structure of its various parts – the nuclear pores, nuclear lamina and nucleolus – should be encoded by host-cell genes. That's not the case. All of them are composed of a chimeric mixture of proteins, some encoded by bacterial genes, a few by archaeal genes, and the rest by genes found only in eukaryotes. It's practically impossible to explain this pattern unless the nucleus evolved *after* the acquisition of mitochondria, in the wake of unruly gene transfers. It's often said that in the evolution of the eukaryotic cell, the endosymbionts were transformed, almost (but not quite) beyond recognition, into mitochondria. It's less appreciated that the host cell underwent an even more dramatic makeover. It started out as a simple archaeon, and acquired endosymbionts. These bombarded their unwitting host with DNA and introns, driving the evolution of the nucleus. And not just the nucleus: sex went hand in hand.

The origin of sex

We've noted that sex arose very early in eukaryotic evolution. I implied, too, that the origins of sex might have had something to do with intron bombardment. How so? Let's first quickly recap what we are trying to explain.

True sex, as practised by eukaryotes, involves the fusion of two gametes (for us, the sperm and egg), each gamete having half the normal quota of chromosomes. You and I are diploid, along with most other multicellular eukaryotes. That means we have two copies of each of our genes, one from each parent. More specifically, we have two copies of each chromosome, known as sister chromosomes. Iconic images of chromosomes might make them look like unchanging physical structures, but that's far from the case. During the formation of gametes, the chromosomes are *recombined*, fusing bits of one with pieces of another, giving new combinations of genes that have most likely never been seen before (**Figure 28**). Work your way down a newly recombined chromosome, gene by gene, and you'll find some genes from your mother, some from your father. The chromosomes are now separated in the process of meiosis (literally, 'reductive cell division') to form haploid gametes, each with a single copy of every chromosome. Two gametes, each with recombined chromosomes, finally fuse to form the fertilised egg, a new individual with a unique combination of genes – your child.

The problem with the origin of sex is not that a lot of new machinery had to evolve. Recombination works by lining up the two sister chromosomes side by side. Sections of one chromosome are then physically transferred on to its sister, and vice versa, by way of cross-over points. This physical lining up of chromosomes and recombination of genes also occurs in bacteria and archaea during lateral gene transfer, but is not usually reciprocal: it is used to repair damaged chromosomes or to reload genes that had been deleted from the chromosome. The molecular machinery is basically the same; what differs in sex is the scope and the reciprocality. Sex is a reciprocal recombination across the entire genome. That entails the fusion of whole cells, and the physical transfer of entire genomes, which is rarely if ever seen in prokaryotes.

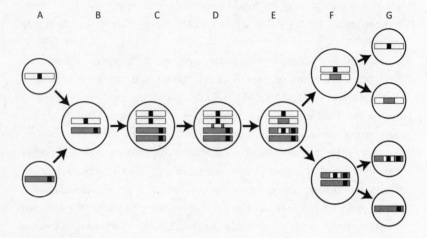

Figure 28 **Sex and recombination in eukaryotes**

A simplified depiction of the sexual cycle: the fusion of two gametes followed by a two-step meiosis with recombination to generate new, genetically distinct gametes. In **A** two gametes with a single copy of an equivalent (but genetically distinct) chromosome fuse together to form a zygote with two copies of the chromosome **B**. Notice the black bars, which could signify either a harmful mutation, or a beneficial variant of specific genes. In the first step of meiosis **C** the chromosomes are aligned and then duplicated, to give four equivalent copies. Two or more of these chromosomes are then recombined **D**. Sections of DNA are reciprocally crossed over from one chromosome to another, to fashion new chromosomes containing bits of the original paternal and bits of the original maternal chromosome **E**. Two rounds of reductive cell division separate these chromosomes to give **F** and finally a new selection gametes **G**. Notice that two of these gametes are identical to the original gametes, but two now differ. If the black bar signifies a harmful mutation, sex has here generated one gamete with no mutations, and one gamete with two; the latter can be eliminated by selection. Conversely, if the black bar signifies a beneficial variant, then sex has united both of them in a single gamete, allowing selection to favour both simultaneously. In short, sex increases the variance (the difference) between gametes, making them more visible to selection, so eliminating mutations and favouring beneficial variants over time.

Sex was considered the 'queen' of biological problems in the twentieth century, but we now have a good appreciation of why it helps, at least in relation to strict asexual reproduction (cloning). Sex breaks up rigid combinations of genes, allowing natural selection to 'see' individual genes, to parse all our qualities one by one. That helps in fending off debilitating parasites, as well as adapting to changing environments, and maintaining necessary variation in a population. Just as medieval stonemasons once carved the back of sculptures that are hidden in the recesses of cathedrals, because they were still visible to God, so sex allows the all-seeing eye of natural selection to inspect her works, gene by gene. Sex gives us 'fluid' chromosomes, ever-changing combinations of genes (technically *alleles*[3]), which allows natural selection to discriminate between organisms with unprecedented finesse.

Imagine 100 genes lined up on a chromosome that never recombines. Selection can only ever discriminate the fitness of the whole chromosome. Let's say there are a few really critical genes on this chromosome – any mutations in them would almost always result in death. Critically, however, mutations on less critical genes become nearly invisible to selection. Slightly harmful mutations can accumulate in these genes, as their negative effects are offset by the big benefits of the few critical genes. As a result, the fitness of the chromosome, and the individual, is gradually undermined. This is roughly what happens to the Y chromosome in men – the lack of recombination means that most genes are in a state of slow degeneration; only the critical genes can be preserved by selection. In the end, the entire chromosome can be lost, as indeed has happened in the mole vole *Ellobius lutescens*.

---

3 Variants of the same gene are termed 'alleles'. Specific genes remain in the same position on a chromosome, the 'locus', but the actual sequence of a specific gene can vary between individuals. If particular variants are common in a population, they are known as alleles. Alleles are polymorphic variants of the same gene, at the same locus. They differ from mutants in frequency. New mutations are present at a low frequency in a population. If they offer an advantage, they may spread through the population until this advantage is counterbalanced by some disadvantage. They have become alleles.

But it's even worse if selection acts positively. Consider what happens if a rare positive mutation in a critical gene is so beneficial that it sweeps through the population. Organisms that inherit the new mutation dominate, and the gene ultimately spreads to 'fixation': all organisms in the population end up with a copy of the gene. But natural selection can only 'see' the whole chromosome. This means that the other 99 genes on the chromosome also become fixed in the population – they go along for the ride and are said to 'hitch-hike' to fixation. This is a disaster. Imagine there are two or three versions (alleles) of each gene in the population. That gives between 10,000 and 1 million different possible combinations of alleles. After fixation, all this variation is wiped out, leaving the entire population with a single combination of the 100 genes – those that happened to share the chromosome with the recently fixed gene – a catastrophic loss of variation. And, of course, a mere 100 genes is a gross oversimplification: asexual organisms have many thousands of genes, all of which are purged of variation in a single selective sweep. The 'effective' population size is hugely diminished, making asexual populations far more vulnerable to extinction.[4] That's exactly what does happen to most asexuals – almost all clonal plants and animals fall extinct within a few million years.

These two processes – accumulation of mildly damaging mutations, and loss of variation in selective sweeps – are together known as *selective interference*. Without recombination, selection on certain genes interferes with selection on others. By generating chromosomes with different combinations of alleles – 'fluid chromosomes' – sex allows selection to act on all genes individually. Selection, like God, can now see all our vices and virtues, gene by gene. That's the great advantage of sex.

---

4 The effective population size reflects the amount of genetic variation in a population. In terms of a parasitic infection, a clonal population might as well be a single individual, as any parasitic adaptation that allows it to target a particular gene combination could tear through the entire population. Conversely, large sexual populations tend to have a lot of genetic variation in alleles (while all sharing the same genes). That variation means that some organisms are likely to be resistant to this particular parasitic infection. The effective population size is greater, even if the number of individuals is the same.

But there are also serious disadvantages to sex, hence its long standing as the queen of evolutionary problems. Sex breaks up combinations of alleles proved to be successful in a particular environment, randomising the very genes that helped our parents to thrive. The gene pack is shuffled again every generation, with never a chance to clone an exact copy of a genius, another Mozart. Worse, there is the 'twofold cost of sex'. When a clonal cell divides, it produces two daughter cells, each of which goes on to produce another two daughters, and so on. The growth of a population is exponential. If a sexual cell produces two daughter cells, these must fuse with each other to form a new individual that can produce two more daughter cells. So an asexual population doubles in size each generation, whereas a sexual population remains the same size. And compared with just cloning a nice copy of yourself, sex introduces the problem of finding a mate, with all its emotional (and financial) costs. And there's the cost of males. Clone yourself and there's no need for all those aggressive, prancing males, locking horns, fanning tails or dominating boardrooms. And we'd be rid of horrible sexually transmitted diseases like AIDS or syphilis, and the opportunity for genetic freeloaders – viruses and 'jumping genes' – to riddle our genomes with junk.

The puzzle is that sex is ubiquitous among eukaryotes. One might think that the advantages would offset the costs under certain circumstances but not others. To a point this is true, in that microbes may divide asexually for 30 generations or so, before indulging in occasional sex, typically when in a state of stress. But sex is far more widespread than seems reasonable. This is probably because the last common ancestor of eukaryotes was already sexual, and hence all her descendants were sexual too. While many micro-organisms no longer have regular sex, very few ever lost sex altogether without falling extinct. The costs of *never* having sex are therefore high. A similar argument should apply to the earliest eukaryotes. Those that never had sex – arguably all those that had not 'invented' sex – were likely to fall extinct.

But here we run again into the problem of lateral gene transfer, which is similar to sex in that it recombines genes, producing 'fluid chromosomes'. Until recently, bacteria were perceived as the grand masters of cloning.

They grow at exponential rates. If totally unconstrained, a single *E. coli* bacterium, doubling every 30 minutes, would produce a colony with the mass of the earth in three days flat. As it happens, though, *E. coli* can do much more than that. They can also swap their genes around, incorporating new genes on to their chromosomes by lateral gene transfer while losing other unwanted genes. The bacteria that give you gastric flu may differ in 30% of their genes compared with the same 'species' up your nose. So bacteria enjoy the benefits of sex (fluid chromosomes) along with the speed and simplicity of cloning. But they don't fuse whole cells together, and they don't have two sexes, and so they avoid many of the disadvantages of sex. They would seem to have the best of both worlds. So why did sex arise from lateral gene transfer in the earliest eukaryotes?

Work from the mathematical population geneticists Sally Otto and Nick Barton points to an unholy trinity of factors that conspicuously relates to circumstances at the origin of eukaryotes: the benefits of sex are greatest when the mutation rate is high, selection pressure is strong, and there is a lot of variation in a population.

Take the mutation rate first. With asexual reproduction, a high mutation rate increases the rate of accumulation of mildly damaging mutations, and also the loss of variation from selective sweeps: it increases the severity of selective interference. Given an early intron invasion, the first eukaryotes must have had a high mutation rate. How high exactly is hard to constrain, but it might be possible to do so by modelling. I'm working on this question with Andrew Pomiankowski and Jez Owen, a PhD student with a background in physics and an interest in these big questions in biology. Jez is right now developing a computational model to figure out where sex scores over lateral gene transfer. There's a second factor to consider here too – genome size. Even if the mutation rate remains the same (let's say one lethal mutation in every 10 billion DNA letters), it is not possible to expand a genome indefinitely without some sort of a mutational meltdown. In this case, cells with a genome of less than 10 billion letters would be fine, but cells with a genome much larger than that would die, as they would all suffer a lethal mutation. The acquisition of mitochondria at the origin of eukaryotes must have exacerbated both problems – they almost certainly increased the mutation rate, and they enabled a massive expansion of genome size, over several orders of magnitude.

It might well be that sex was the only solution to this problem. While lateral gene transfer could in principle avoid selective interference through recombination, Jez's work suggests that this can only go so far. The larger the genome, the harder it becomes to pick up the 'correct' gene by lateral gene transfer; that's really just a numbers game. The only way to ensure that a genome has all the genes it needs, in full working order, is to retain all of them, and to recombine them regularly across the entire genome. That can't be achieved by lateral gene transfer – it needs sex, 'total sex', involving recombination across the whole genome.

What about the strength of selection? Again, introns may be important. In modern organisms, the classic selection pressures favouring sex are parasitic infections and variable environments. Even then, selection has to be strong for sex to be better than cloning – for example, parasites must be common and debilitating to favour sex. No doubt these same factors applied to early eukaryotes too, but they also had to contend with a debilitating early intron invasion – parasitic genes. Why would mobile introns drive the evolution of sex? Because genome-wide recombination increases variance, forming some cells with introns in damaging places, and other cells with introns in less hazardous places. Selection then acts to weed out the worst cells. Lateral gene transfer is piecemeal, and cannot produce systematic variation, in which some cells have their genes cleaned up, while others accumulate more than their share of mutations. In his brilliant book *Mendel's Demon*, Mark Ridley compared sex with the New Testament view of sin – just as Christ died for the collected sins of humanity, so too sex can bring together the accumulated mutations of a population into a single scapegoat, and then crucify it.

The amount of variation between cells could also relate to introns. Both archaea and bacteria usually have a single circular chromosome, whereas eukaryotes have multiple straight chromosomes. Why? A simple answer is that introns can cause errors as they splice themselves in and out of the genome. If they fail to rejoin the two ends of a chromosome after cutting themselves out, that leaves a break in the chromosome. A single break in a circular chromosome gives a straight chromosome; several breaks give several straight chromosomes. So recombinatorial errors produced by mobile introns could have produced multiple straight chromosomes in the early eukaryotes.

That must have given early eukaryotes terrible problems with their cell cycle. Different cells would have had different numbers of chromosomes, each accumulating different mutations or deletions. They would also have been picking up new genes and DNA from their mitochondria. Copying errors would no doubt duplicate chromosomes. It's hard to see what lateral gene transfer could contribute in this context. But standard bacterial recombination – lining up chromosomes, loading up missing genes – would ensure that cells tended to accumulate genes and traits. Only sex could accumulate genes that worked, and be rid of those that didn't. This tendency to pick up new genes and DNA by sex and recombination easily accounts for the swelling of early eukaryotic genomes. Accumulating genes in this way must have solved some of the problems of genetic instability, while the energetic advantages of having mitochondria meant that, unlike bacteria, there was no energetic penalty. All this is speculative, to be sure, but the possibilities can be constrained by mathematical modelling.

How did cells physically segregate their chromosomes? The answer may lie in the machinery used by bacteria to separate large plasmids – mobile 'cassettes' of genes that encode traits such as antibiotic resistance. Large plasmids are typically segregated in bacterial division on a scaffold of microtubules that resembles the spindle used by eukaryotes. It's plausible that the plasmid segregation machinery was requisitioned by early eukaryotes to separate their varied chromosomes. It is not only plasmids that are segregated in this way – some bacterial species seem to separate their chromosomes on relatively dynamic spindles, rather than using the cell membrane as normal. It may be that better sampling of the prokaryotic world will give us more clues about the physical origins of eukaryotic chromosomal segregation in mitosis and meioisis.

It's almost unknown among bacteria with cell walls, though some archaea are known to fuse. The loss of the cell wall would certainly have made fusion far more likely; and L-form bacteria, which have lost their cell wall, do indeed fuse with each other quite readily. The number of controls over cell fusion in modern eukaryotes also implies it might have been hard to stop their ancestors fusing together. Early fusions could even have been promoted by mitochondria, as argued by the ingenious evolutionary biologist Neil Blackstone. Think about their predicament. As endosymbionts, they could not leave their host cells and simply infect another one, so their own evolutionary success was tied to the growth of their hosts. If their hosts were crippled by mutations and unable to grow, the mitochondria would be stuck too, unable to proliferate themselves. But what if they could somehow induce fusion with another cell? This is a win-win situation. The host cell acquires a complementary genome, thereby enabling recombination, or perhaps simply masking mutations on particular genes with potentially clean copies of the same genes – the benefits of outbreeding. Because cell fusion permitted renewed growth of the host cell, the mitochondria could revert to copying themselves too. So early mitochondria could have been agitating for sex![5] That might have solved their immediate problem, but ironically, it only opened the door to another, even more pervasive, issue: competition between mitochondria. The solution might just have been that other puzzling aspect of sex – the evolution of two sexes.

Two sexes

'No practical biologist interested in sexual reproduction would be led to work out the detailed consequences experienced by organisms having three

---

5 Blackstone has even suggested a possible mechanism that derives from the biophysics of mitochondria. Host cells whose growth is crippled by mutations would have low ATP demands, so they would break little ATP back down into ADP. Because electron flow in respiration depends on ADP concentration, the respiratory chain would tend to fill up with electrons and become more reactive, forming oxygen free radicals (more on this in the next chapter). In some algae today, free-radical leakage from mitochondria induces the formation of gametes and sex; and this response can be blocked by giving them antioxidants. Could free radicals have triggered membrane fusion directly? It's possible. Radiation damage is known to cause membrane fusion through a free-radical mechanism. If so, a natural biophysical process could have served as the basis for subsequent natural selection.

or more sexes; yet what else should he do if he wishes to understand why the sexes are, in fact, always two?' So said Sir Ronald Fisher, one of the founding fathers of evolutionary genetics. The problem has yet to be conclusively solved.

On paper, two sexes seem to be the worst of all possible worlds. Imagine if everyone were the same sex – we could all mate with each other. We would double our choice of partners in one go. Surely that would make everything easier! If, for some reason, we are obliged to have more than one sex, then three or four sexes ought to be better than two. Even if restricted to mating with other sexes, we could then couple with two-thirds or three-quarters of the population rather than a mere half. It would still take two partners, of course, but there's no obvious reason why these partners couldn't be the same sex, or multiple sexes, or for that matter hermaphrodites. The practical difficulties with hermaphrodites gives away part of the problem: neither partner wants to bear the cost of being the 'female'. Hermaphroditic species such as flatworms go to bizarre lengths to avoid being inseminated, fighting pitched battles with their penises, their semen burning gaping holes in the vanquished. This is lively natural history, but it is circular as an argument, as it takes for granted that there are greater biological costs to being female. Why should there be? What actually is the difference between male and female? The split runs deep and has nothing to do with X and Y chromosomes, or even with egg cells and sperm. Two sexes, or at least mating types, are also found in single-celled eukaryotes, such as some algae and fungi. Their gametes are microscopic and the two sexes look indistinguishable but they're still as discriminating as you and me.

One of the deepest distinctions between the two sexes relates to the inheritance of mitochondria – one sex passes on its mitochondria, while the other sex does not. This distinction applies equally to humans (all our mitochondria come from our mother, 100,000 of them packed into the egg) and to algae such as *Chlamydomonas*. Even though such algae produce identical gametes (or isogametes) only one sex passes on its mitochondria; the other suffers the indignity of having its mitochondria digested from within. In fact it's specifically the mitochondrial DNA that gets digested; the problem seems to be the mitochondrial genes, not the morphological structure. So we have a very peculiar situation, in which mitochondria apparently agitate for sex, as we've just seen, but the outcome is not that they spread from cell to cell but that half of them get digested. What's going on here?

The most graphic possibility is selfish conflict. There is no real competition between cells that are all genetically the same. That's how our own cells are tamed, so that they cooperate together to form our bodies. All our cells are genetically identical; we are giant clones. But genetically different cells do compete, with some mutants (cells with genetic changes) producing cancer; and much the same happens if genetically different mitochondria mix in the same cell. Those cells or mitochondria that replicate the fastest will tend to prevail, even if that is detrimental to the host organism, producing a kind of mitochondrial cancer. That's because cells are autonomous self-replicating entities in their own right, and they are always poised to grow and divide if they can. French Nobel laureate François Jacob once said that the dream of every cell is to become two cells. The surprise is not that they often do, but that they can be restrained for long enough to make a human being. For these reasons, mixing two populations of mitochondria in the same cell is just asking for trouble.

This idea goes back several decades, and comes with the seal of some of the greatest evolutionary biologists, including Bill Hamilton. But the idea is not beyond challenge. For a start, there are known exceptions, in which mitochondria mix freely, and it does not always end in disaster. Then there is a practical problem. Imagine a mitochondrial mutation that gives a replicative advantage. The mutant mitochondria outgrow the rest. Either this is lethal, in which case the mutants will die out along with the host cells, or it is not, and the mutants spread through the population. Any genetic constraint on their spread (for example, some change in a nuclear gene that prevents mitochondrial mixing) has to arise quickly, to catch the mutant in the act of spreading. If just the right gene does *not* arise in time, it's too late. Nothing is gained if the mutant has already spread to fixation. Evolution is blind and has no foresight. It can't anticipate the next mitochondrial mutant. And there's a third point that makes me suspect that fast-replicating mitochondria are not as bad as all that – the fact that mitochondria have retained so few genes. There may be many reasons for this, but selection on mitochondria for fast replication is surely among them. That implies there have been numerous mutations that sped up mitochondrial replication over time. They were not eliminated by the evolution of two sexes.

For these reasons, I suggested a new idea in an earlier book: perhaps the problem relates rather to the requirement for mitochondrial genes to adapt to genes in the nucleus. I'll say more about this in the next chapter. For now, let's just note the key point: for respiration to work properly, genes in the mitochondria and the nucleus need to cooperate with each other, and mutations in either genome can undermine physical fitness. I proposed that uniparental inheritance, in which only one sex passes on the mitochondria, might improve the coadaptation of the two genomes. The idea makes reasonable sense to me, but there it would have rested had not Zena Hadjivasiliou, an able mathematician with a budding interest in biology, embarked on a PhD with me and Andrew Pomiankowski.

Zena did indeed show that uniparental inheritance improves the coadaptation of the mitochondrial and nuclear genomes. The reason is simple enough and relates to the effects of sampling, a theme that will return with intriguing variations. Imagine a cell with 100 genetically different mitochondria. You remove one of them, place it all by itself inside another cell, and then copy it, until you have 100 mitochondria again. Barring a few new mutations, these mitochondria will all be the same. Clones. Now do the same with the next mitochondrion and keep going until you have copied all 100. Each of your 100 new cells will have different populations of mitochondria, some of them good, some bad. You have increased the *variance* between these cells. If you had just copied the whole cell 100 times, each daughter cell would have had roughly the same mix of mitochondria as the parent cell. Natural selection would not be able to distinguish between them – they are all too similar. But by sampling and cloning the sample, you produced a range of cells, some of them fitter than the original, others less fit.

This is an extreme example, but illustrates the point of uniparental inheritance. By sampling a few mitochondria from only one of the two parents, uniparental inheritance increases the variance of mitochondria between fertilised egg cells. This greater variety is more visible to natural selection, which can eliminate the worst cells, leaving the better cells behind. The fitness of the population improves over generations. Intriguingly, this is practically the same advantage as sex itself, but sex increases the variance of nuclear genes, whereas two sexes increase the variance of mitochondria between cells. It's as simple as that. Or so we thought.

Our study was a straight comparison of fitness with and without uniparental inheritance, but at this point we had not considered what would happen if a gene imposing uniparental inheritance were to arise in a population of biparental cells, in which both gametes pass on the mitochondria. Would it spread to fixation? If so, we would have evolved two sexes: one sex would pass on its mitochondria and the other sex would have its mitochondria killed. We developed our model to test this possibility. For good measure we compared our coadaptation hypothesis with the outcomes arising from selfish conflict, as discussed above, and a simple accumulation of mutations.[6] The results came as a surprise, and at least initially were disappointing. The gene would not spread, certainly not to fixation.

The problem was that the fitness costs depend on the number of mutant mitochondria: the more mutants, the higher the costs. Conversely, the benefits of uniparental inheritance also depend on the mutation load, but this time the other way round: the smaller the mutant load, the lower the benefit. In other words the costs and benefits of uniparental inheritance are not fixed, but change with the number of mutants in the population; and that can be lowered by just a few rounds of uniparental inheritance (**Figure 29**). We found that uniparental inheritance did indeed improve the fitness of a population in all three models, but as the gene for uniparental inheritance begins to spread through a population, its benefits dwindle until they are offset by

---

6 From a mathematical point of view, all three theories turned out to be variants of each other: each one depends on the mutation rate. In a simple mutation model, the rate of accumulation of mutants obviously depends on the mutation rate. Likewise, when a selfish mutant arises, it replicates a little faster than the wild type, meaning that the new mutant spreads through the population. Mathematically that equates to a faster mutation rate, which is to say, there are more mutants in a given time. The coadaptation model does the opposite. The effective mutation rate is lowered because nuclear genes can adapt to mitochondrial mutants, meaning they are no longer detrimental, hence by our definition they're not mutants.

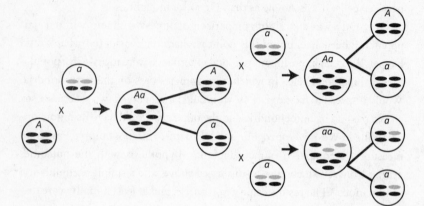

Figure 29 **The 'leakage' of fitness benefits in mitochondrial inheritance**
A and *a* are gametes with different versions (alleles) of a particular gene in the
nucleus, denoted *A* and *a*. Gametes with *a* pass on their mitochondria when they
fuse with another *a* gamete. Gametes with *A* are 'uniparental mutants': if an *A*
gamete fuses with an *a* gamete, only the *A* gamete passes on its mitochondria.
The first mating here shows a fusion of *A* and *a* gametes, to produce a zygote
with both nuclear alleles (*Aa*), but all mitochondria deriving from *A*. If *a* contains
some defective mitochondria (pale shade) these are eliminated by uniparental
inheritance. The zygote now produces two gametes, one with the *A* allele, and
one with the *a* allele. Each of these fuses with an *a* gamete containing defective
mitochondria (pale shade). In the upper cross, *A* and *a* gametes generate an *Aa*
zygote, with all mitochondria deriving from the *A* gamete, thereby eliminating
the defective (pale) mitochondria. In the lower cross, two *a* gametes fuse, and the
defective mitochondria are passed on to the *aa* zygote. Each of these zygotes (*Aa*
and *aa*) now forms gametes. The *a* mitochondria have now been 'cleaned up' by a
couple of rounds of uniparental inheritance. That improves the fitness of
biparental gametes, so the fitness benefit 'leaks' through the population, ultimately
arresting its own spread.

the disadvantages – the main disadvantage being that uniparental cells mate with a smaller fraction of the population. The trade-off reaches equilibrium when barely 20% of the population is uniparental. High mutation rates could force it up to 50% of the population; but the other half of the population could continue to mate among themselves, giving, if anything, three sexes. The bottom line is that mitochondrial inheritance will not drive the evolution of two mating types. Uniparental inheritance increases the variance between gametes, improving fitness, but this benefit isn't strong enough by itself to drive the evolution of mating types.

Well, this was a direct disproof of my own idea, so I didn't like it much. We tried everything we could think of to make it work, but eventually I had to concede that there are no realistic circumstances in which a uniparental mutant could drive the evolution of two mating types. Mating types must have evolved for some other reason.[7] Even so, uniparental inheritance exists. Our model would simply be wrong if we could not explain that. In fact, we showed that if two mating types *did* already exist, for some other reason, then certain conditions *could* fix uniparental inheritance: specifically, a large number of mitochondria, and a high mitochondrial mutation rate. Our conclusion seemed incontrovertible; and our explanation sits more comfortably with the known exceptions to uniparental inheritance in the natural world. It also made sense of the fact that uniparental inheritance is practically universal among multicellular organisms, animals such as ourselves, which do in general have large numbers of mitochondria and high mutation rates.

This is a fine example of why mathematical population genetics is

---

7 There are plenty of other possibilities, ranging from ensuring outbreeding to signalling and pheromones. Given that two cells fuse in sex, they first have to find each other, and make sure they fuse with the right cell – another cell of the same species. Cells typically find each other by 'chemotaxis', which is to say they produce a pheromone, in effect a 'smell', and they move towards the source of the smell, up a concentration gradient. If both gametes produce the same pheromone, they can confuse themselves. They're likely to swim around in small circles, smelling their own pheromone. It is generally better for only one gamete to produce a pheromone and the other to swim towards it, so the distinction between mating types could relate to the problem of finding a mate.

important: hypotheses need to be tested formally, by whatever methods are possible; in this case, a formal model showed clearly that uniparental inheritance cannot fix in a population unless two mating types already exist. This is as close to rigorous proof as we can get. But all was not yet lost. The difference between mating types and 'true' sexes (in which males and females are obviously different) is opaque. Many plants and algae have both mating types and sexes. Perhaps our definition of sexes was wrong, and we really should be considering the evolution of true sexes, rather than two ostensibly identical mating types. Could uniparental inheritance account for the distinction between true sexes in animals and plants? If so, mating types might have arisen for other reasons, but the evolution of true sexes could still have been driven by mitochondrial inheritance. Frankly, that seemed a weak idea, but worth a look. That reasoning did not begin to prepare us for the revelatory answer that we actually found, an answer that emerged precisely because we did *not* start out with the normal assumption that uniparental inheritance is universal, but with the disappointing conclusions of our own previous study.

Immortal germline, mortal body

Animals have large numbers of mitochondria, and we use them ceaselessly to power our supercharged lifestyles, giving us high mitochondrial mutation rates, right? More or less right. We have hundreds or thousands of mitochondria in each cell. We don't know their mutation rate for sure (it's difficult to measure directly) but we do know that over many generations, our mitochondrial genes evolve some 10–50 times faster than genes in the nucleus. This implies that uniparental inheritance ought to fix readily in animals. In our model, we did indeed find that uniparental inheritance will fix more easily in multicellular than single-celled organisms. No surprises there.

But we are easily misled by thinking about ourselves. The first animals were not like us: they were more like sponges or corals, sessile filter feeders that don't move around, at least not in their adult forms. Not surprisingly, they don't have many mitochondria, and the mitochondrial mutation rate is low – lower, if anything, than in the nuclear genes. This was the starting point for the PhD student Arunas Radzvilavicius, yet another gifted physicist attracted to the big problems of biology. One begins to wonder if all the most interesting problems in physics are now in biology.

What Arunas realised is that simple cell division in multicellular organisms has a rather similar effect to uniparental inheritance: it increases the variance between cells. Why? Each round of cell division apportions the mitochondrial population randomly between the daughter cells. If there are a few mutants, the chances of them being distributed exactly equally is low – it's far more likely that one daughter cell will receive a few more mutants than the other. If this is repeated over many rounds of cell division, the outcome is greater variance; some great-great-great-granddaughter cells will end up inheriting a greater mutant load than others. Whether this is a good or a bad thing depends on which cells receive the bad mitochondria, and how many there are.

Imagine an organism like a sponge, in which all the cells are quite similar. It's not differentiated into lots of specialised tissues, like brain and intestine. Cut up a living sponge into small pieces (don't do this at home) and it can regenerate itself from those bits and pieces. It can do so because stem cells, lurking more or less anywhere, can give rise to new germ cells as well as new somatic (body) cells. In this regard, sponges are similar to plants – neither of them sequesters a specialised germline early in development, but instead they generate gametes from stem cells in many tissues. This difference is critical. We have a dedicated germline, which is hidden away early on in embryonic development. A mammal would normally never produce germ cells from stem cells in the liver. Sponges, corals and plants, however, can grow new sexual organs producing gametes from many different places. There are explanations for these differences, rooted in competition between cells, but they are not really compelling.[8] What Arunas found is that all

---

8 The developmental biologist Leo Buss has argued, for example, that animal cells, being mobile, are more likely to invade the germline, in a selfish attempt to perpetrate themselves, than plant cells, whose cumbersome cell wall renders them virtually immobile. But is that true of corals and sponges too, which are composed of perfectly

these organisms have one thing in common: they have a small number of mitochondria and a low mitochondrial mutation rate. And the few mutations that do occur can be eliminated by *segregation*. It works like this.

Recall that multiple rounds of cell division increase the variance between cells. That goes for germ cells too. If the germ cells are sequestered early on in development, there can't be much difference between them – the few rounds of cell division don't generate much variance. But if germ cells are selected at random from adult tissues, then there will be much greater differences between them (**Figure 30**). Multiple rounds of cell division mean that some germ cells accumulate more mutations than others. Some will be nearly perfect, others a dreadful mess – there is high variance between them. That is what natural selection needs: it can weed out all the bad cells, so only the good ones survive. Over generations, the quality of germ cells increases; selecting them randomly from adult tissues works better than hiding them away, putting them 'on ice' early in development.

So greater variance is good for the germline, but it can be devastating for the health of an adult. Bad germ cells are eliminated by selection, leaving the better ones to seed the next generation; but what about bad stem cells, which give rise to new adult tissues? These will tend to produce dysfunctional tissues that may be unable to support the organism. The fitness of the organism as a whole depends on the fitness of its worst organ. If I have a heart attack, the function of my kidneys is immaterial: my healthy organs will die along with the rest of me. So there are both advantages and disadvantages to increasing mitochondrial variance in an organism, and the advantage to the germline may well be offset by the disadvantage to the body as a whole. The degree to which it is offset depends on the number of tissues and the mutation rate.

The more tissues there are in an adult, the greater the likelihood that a vital tissue will accumulate all the worst mitochondria. Conversely, with only one tissue type, this is not a problem, as there is no interdependence – no organs whose failure can undermine the function of the whole individual. In the case of a simple organism with a single tissue, then, increased

---

mobile animal cells? I doubt it. Yet they have no more of a germline than do plants.

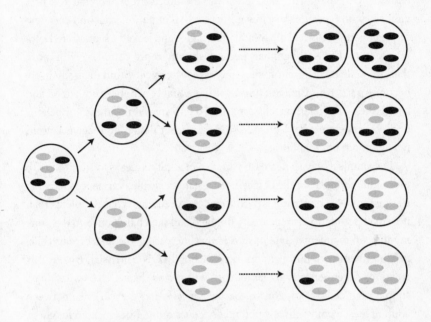

Figure 30 **Random segregation increases variance between cells**
If a cell starts with a mixture of different types of mitochondria, which are
doubled and then divided roughly equally between two daughter cells, the
proportions will vary slightly with each cell division. Over time these differences
are amplified, as each cell partitions an increasingly distinct population of
mitochondria. If the final daughter cells at the right become gametes, then
repeated cell division has the effect of increasing variance between gametes.
Some of these gametes are very good, and others very bad, increasing the
visibility to natural selection: exactly the same effect as uniparental inheritance,
and a Good Thing. Conversely, if the cells at the right are progenitor cells that
give rise to a new tissue or organ, then this increased variance is a disaster. Now
some tissues will function well but others will fail, undermining the fitness of the
organism as a whole. One way to lower the variance between tissue progenitor
cells is to increase the number of mitochondria in the zygote, such that the
number of mitochondria partitioned initially is much greater. This can be
achieved by increasing the size of the egg cell, giving rise to 'anisogamy' (large
egg, small sperm).

variance is unequivocally good: it's beneficial for the germline and not par-ticularly detrimental to the body. We therefore predicted that the first animals, with (presumably) low mitochondrial mutation rates and very few tissues, should have had biparental inheritance and lacked a sequestered ger-mline. But when early animals became slightly more complex, with more than a couple of different tissues, increased variance within the body itself becomes disastrous for adult fitness, as it inevitably produces both good and bad tissues – the heart attack scenario. To improve adult fitness, mitochon-drial variance must be decreased so that nascent tissues all receive similar, mostly good, mitochondria.

The simplest way to decrease variance in adult tissues is to start out with more mitochondria in the egg cell. As a statistical rule, variance is lower if a large founder population is partitioned between numerous recipients than if a small population is repeatedly doubled and then partitioned to the same number of recipients. The upshot is that increasing the size of egg cells, packing them with more and more mitochondria, is beneficial. By our cal-culations, a gene specifying larger eggs will spread through a population of simple multicellular organisms, because it *decreases* the variance between adult tissues, ironing out any potentially devastating differences in function. On the other hand, less variance is not good for gametes, which become more similar to each other, and so less 'visible' to natural selection. How can these two opposing tendencies be reconciled? Simple! If only one of the two gametes, the egg cell, increases in size, whereas the other shrinks, becoming sperm, that solves both problems. The large egg cell decreases the variance between tissues, improving adult fitness, whereas the exclusion of mito-chondria from sperm ultimately results in uniparental inheritance, with only one parent passing on its mitochondria. We've already noted that uniparen-tal inheritance of mitochondria increases the variance between gametes, so improving their fitness. In other words, from the simplest of starting points, both anisogamy (distinct gametes, sperm and egg) followed by uniparental inheritance will tend to evolve in organisms with more than one tissue.

I have to stress that all this assumes a low mitochondrial mutation rate. That's known to be the case in sponges, corals and plants, but it is not the case in 'higher' animals. What happens if the mutation rate rises? The benefit of delaying germ-cell production is now lost. Our model shows that mutations accumulate quickly, leaving late germ cells riddled with mutations. As the geneticist James Crow put it, the greatest mutational health hazard in the population is fertile old men. Thankfully, uniparental inheritance means that men don't pass on their mitochondria at all. Given a faster mutation rate, we find that a gene which induces early sequestration of the germline will spread through a population: hiving off an early germline, putting female gametes on ice, limits the accumulation of mitochondrial mutations. Adaptations that specifically lower the germline mutation rate should also be favoured. In fact, mitochondria in the female germline seem to be switched off, hidden in the primordial egg cells that are sequestered early in the embryonic development of ovaries, as shown by my colleague John Allen. He has long argued that the mitochondria in egg cells are genetic 'templates', which, being inactive, have a lower mutation rate. Our model supports these ideas for modern fast-living animals with numerous mitochondria and rapid mutation rates, but not for their slower-living ancestors, or for wider groups such as plants, algae and protists.

What does all this mean? It means, astonishingly, that mitochondrial variation *alone* can explain the evolution of multicellular organisms that have anisogamy (sperm and eggs), uniparental inheritance, and a germline, in which female germ cells are sequestered early in development – which together form the basis for all sexual differences between males and females. In other words, the inheritance of mitochondria can account for most of the real physical differences between the two sexes. Selfish conflict between cells may play a role too, but is not necessary: the evolution of the germline–soma distinction can be explained without reference to selfish conflict. Critically, our model specifies an order of events that is not what I would have guessed at the outset. I had imagined that uniparental inheritance was the ancestral state, that the germline evolved next, and that the evolution of sperm and egg cells was connected with the divergence of true sexes. Instead, our model implies that the ancestral state was biparental; anisogamy (sperm and egg) arose next, then uniparental inheritance, and finally the germline. Is this revised order correct? There's little information either way. But it is an explicit prediction that can be tested, and we hope to do so.

The places to look first are sponges and corals. Both groups have sperm and eggs, but lack a sequestered germline. Would they develop one if we selected for a higher mitochondrial mutation rate?

Let's draw to a close with a few implications. Why would the mitochondrial mutation rate rise? An increased turnover of cells and proteins would do it, reflecting physical activity. The oxygenation of the oceans soon before the Cambrian explosion favoured the evolution of active bilateral animals. Their greater activity would have raised their mitochondrial mutation rate (which is measurable in phylogenetic comparisons), and that should have forced the sequestration of a dedicated germline in these animals. This was the origin of the immortal germline and the mortal body – the origin of death as a planned and predetermined end point. The germline is immortal in the sense that germ cells can continue dividing forever. They never age or die. Each generation sequesters a germline early in development, which produces the cells that seed the next generation. Individual gametes may become damaged, but the fact that babies are born young means that germ cells alone retain the potential for immortality seen in organisms like sponges that regenerate themselves from bits and pieces. As soon as this specialised germline is hidden away, the rest of the body can specialise for specific purposes, no longer restrained by the need to retain immortal stem cells in their midst. We see for the first time tissues that can no longer regenerate themselves, such as the brain. The disposable soma. These tissues have a limited lifespan, which depends on how long the organism takes to reproduce itself. That depends on how quickly the animal reaches reproductive maturity, the developmental rate, its anticipated lifespan. We see for the first time a trade-off between sex and death, the roots of ageing. We'll look into that in the next chapter.

This chapter has explored the effects of mitochondria on the eukaryotic cell, some of which were dramatic. Recall the central question: why did all eukaryotes evolve a whole series of shared traits that are never found in bacteria or archaea? In the previous chapter, we saw that prokaryotes are constrained by their cell structure, and specifically the requirement for genes to control respiration. The acquisition of mitochondria transformed the selective landscape for eukaryotes, enabling their expansion in cell volume and genome size over four or five orders of magnitude. That trigger was a rare endosymbiosis between two prokaryotes, not far from a freak accident, but the consequences were both severe and predictable. Severe, because a cell lacking a nucleus is highly vulnerable to a barrage of DNA and genetic parasites (introns) from its own endosymbionts. Predictable, because the response of the host cell at each stage – the evolution of a nucleus, sex, two sexes and a germline – can be understood in terms of classical evolutionary genetics, albeit from an unconventional starting point. Some of the ideas in this chapter may turn out to be wrong, as did my hypothesis on the evolution of two sexes; but in that case a fuller understanding turned out to be far richer than I had imagined, accounting instead for the germline–soma distinction, the origins of sex and death. The underlying logic, excavated through rigorous modelling, is at once beautiful and predictable. Life is likely to follow a similar path to complexity elsewhere.

This view of life's history, a 4-billion-year story, places the mitochondria right at the centre of the evolution of the eukaryotic cell. In recent years, medical research has come to a rather similar view: we now appreciate that mitochondria are instrumental in controlling cell death (apoptosis), cancer, degenerative disease, fertility, and more. But my arguments that mitochondria really are the hub of physiology are prone to make some medical researchers cross; the charge is that I lack a properly balanced perspective. Look at any human cell down a microscope, and you will see a wonderful assembly of working parts, of which the mitochondria are just one, admittedly important, cog. But that is not the view from evolution. The view from evolution sees mitochondria as equal partners in the origin of complex life. All eukaryotic traits – all cell physiology – evolved in the ensuing tug of war between these two partners. That tug of war continues to this day. In the final part of this book, we shall see how this interplay underpins our own health, fertility and longevity.

PART IV

PREDICTIONS

7

THE POWER AND THE GLORY

Christ Pantocrator: Ruler of the World. Even beyond Orthodox iconography, there can be no greater artistic challenge than the portrait of Christ in his 'two natures', at once God and man, the stern but loving judge of all humanity. In his left hand, he may carry the Gospel of John: 'I am the light of the world, who follows me will not wander in the darkness but will have the light of life.' Unsurprisingly, given this sober task, the Pantocrator tends to look rather melancholic. From the artist's point of view, capturing the spirit of God in the face of man is not enough: it must be done in mosaic, inside a dome, high over the altar of a fine cathedral. I can't imagine the skills required to get the perspective just right, to catch the light and shade of a living face, to invest tiny pieces of stone with meaning, each piece oblivious of its place in the grand design, yet each one crucial to the full conception. I do know that marginal errors can destroy the whole effect, giving the Creator a disturbingly comical expression; but when done supremely well, as in Cefalù Cathedral in Sicily, even the least religious will recognise the face of God, an eternal monument to the genius of forgotten human craftsmen.[1]

---

1 Cefalù cathedral was begun in 1131, 40 years after the Normans completed their conquest of Sicily in 1091 (a campaign that had extended over 30 years, beginning in 1061, before their more celebrated conquest of England). The cathedral was built in

I am not about to depart in some unanticipated direction. I'm struck by the appeal of mosaics to the human mind, and by the strikingly parallel importance of mosaics in biology – could there be a subconscious connection between the modularity of proteins and cells, and our sense of aesthetics? Our eyes are composed of millions of photoreceptor cells, rods and cones; each receptor is switched on or off with a ray of light, forming an image as a mosaic. This is reconstructed in our mind's eye as a neuronal mosaic, conjured up from splintered features of the image – brightness, colour, contrast, edge, movement. Mosaics stir our emotions in part because they splinter reality in a similar way to our minds. Cells can do this because they are modular units, living tiles, each one with its own vital place, its own job, 40 trillion pieces making up the wonderful three-dimensional mosaic that is a human being.

Mosaics run even deeper in biochemistry. Consider mitochondria. The great respiratory proteins, which transfer electrons from food to oxygen while pumping protons across the mitochondrial membrane, are mosaics of numerous subunits. The largest, complex I, is composed of 45 separate proteins, each one made up of hundreds of amino acids linked together in a long chain. These complexes are often grouped into larger ensembles, 'supercomplexes', which funnel electrons to oxygen. Thousands of supercomplexes, each one an individual mosaic, adorn the majestic cathedral of the mitochondrion. The quality of these mosaics is vitally important. A comical Pantocrator may be no laughing matter, but tiny errors in the position of individual pieces in the respiratory proteins can carry a burden as terrible as any biblical punishment. If just one amino acid is out of place – a single stone in the full mosaic – the consequences may be a crippling degeneration of the muscles and brain, and an early death: a mitochondrial disease. These genetic conditions are horribly unpredictable in their

---

thanksgiving after King Roger II survived shipwreck off the coast. The wonderful churches and palaces of Norman Sicily combine archetypal Norman architecture with Byzantine mosaics and Arabic cupolas. The Pantocrator at Cefalù was produced by Byzantine craftsmen, and some say is even finer than the famous Pantocrator in the Hagia Sofia, in what was then Constantinople. Either way, it's well worth a visit.

severity and their age of onset, depending on exactly which piece is affected, and how often; but all of them reflect the centrality of mitochondria to the very bone of our existence.

So mitochondria are mosaics, and their quality matters in terms of life and death; but there is more. Like the Pantocrator, the respiratory proteins are unique in having 'two natures', the mitochondrial and the nuclear, and these had better be a match made in heaven. The peculiar arrangement of the respiratory chain – the assembly of proteins that conveys electrons from food to oxygen – is shown in **Figure 31**. Most of the core proteins in the mitochondrial inner membrane, shown in darker shade, are encoded by genes located in the mitochondria themselves. The remaining proteins (lighter shading) are encoded by genes in the nucleus. We have known about this strange state of affairs since the early 1970s, when it first became clear that the mitochondrial genome is so small that it cannot possibly encode most of the proteins found in the mitochondria. The old idea that mitochondria are still independent of their host cells is therefore nonsense. Their ostensible autonomy – they give an eerie impression of replicating themselves whenever they feel like it – is a mirage. The fact is that their function depends on two distinct genomes. They can only grow or function if they are wholly provisioned with proteins encoded by both of these genomes.

Let me ram home just how odd this is. Cell respiration – without which we would die within minutes – depends on mosaic respiratory chains that are composed of proteins encoded by two very different genomes. To reach oxygen, electrons must hop down a respiratory chain from one 'redox centre' to the next. Redox centres typically accept or donate electrons one at a time – these are the stepping stones that we discussed in Chapter 2. The redox centres are embedded deep inside the respiratory proteins, their precise positions depending on the structure of the proteins, hence on the sequence of the genes that encode the proteins, and hence on both the mitochondrial and nuclear genomes. As noted, electrons hop by a process known as quantum tunnelling. They appear and disappear from each centre with a probability that depends on several factors – the tugging power of oxygen (more specifically, the reduction potential of the next redox centre), the

Figure 31 **The mosaic respiratory chain**
Protein structures for complex I (left), complex III (centre left), complex IV (centre right) and the ATP synthase (right), all embedded in the inner mitochondrial membrane. The darker subunits, mostly buried within the membrane, are encoded by genes that are physically located in the mitochondria, whereas the paler subunits, mostly peripheral or outside the membrane, are encoded by genes that reside in the nucleus. These two genomes evolve in dramatically different ways – the mitochondrial genes are passed on asexually from mother to daughter, whereas nuclear genes are recombined by sex every generation; and mitochondrial genes (in animals) also accumulate mutations at up to fifty times the rate of nuclear genes. Despite this propensity to diverge, natural selection can generally eliminate dysfunctional mitochondria, maintaining nearly perfect function over billions of years.

distance between adjacent redox centres, and the occupancy (whether the next centre is already occupied by an electron). The precise distance between redox centres is critical. Quantum tunnelling will only take place over very short distances, less than about 14 Å (recall that an ångström (Å) is about the diameter of an atom). Redox centres spaced further apart might as well be infinitely distant, as the likelihood of electrons hopping between them falls to zero. Within this critical range, the rate of hopping depends on the distance between centres. And that depends on how the two genomes interact with each other.

For each ångström increase in distance between redox centres, the speed of electron transfer falls about 10-fold. Let me say that again. There is a 10-fold decrease in the rate of electron transfer for every 1 Å increase in distance between redox centres! That's roughly the scale of electrical interactions between adjacent atoms, for example the 'hydrogen bonds' between negatively and positively charged amino acids in proteins. If a mutation alters the identity of an amino acid in a protein, hydrogen bonds may be broken, or new ones formed. Whole webs of hydrogen bonds may shift a little, including those that pinion a redox centre into its correct position. It might well move by an ångström or so. The consequences of such tiny shifts are magnified by quantum tunnelling: an ångström this way or that could slow down electron transfer by an order of magnitude, or speed it by an equivalent factor. That's one reason why mitochondrial mutations can be so catastrophic.

This precarious arrangement is exacerbated by the fact that the mitochondrial and nuclear genomes are diverging continuously. In the previous chapter, we saw that the evolution of both sex and two sexes could have been related to the acquisition of mitochondria. Sex is needed to maintain the function of individual genes in large genomes, whereas two sexes help maintain the quality of mitochondria. The unforeseen consequence was that these two genomes evolve in totally different ways. Nuclear genes are recombined by sex every generation, whereas mitochondrial genes pass from mother to daughter in the egg cell, rarely if ever recombining. Even worse, mitochondrial genes evolve 10–50 times faster than genes in the nucleus, in terms of their rate of sequence change over generations, at least in animals. This means that proteins encoded by mitochondrial genes are morphing faster and in different ways compared with proteins encoded by genes in the nucleus; yet still they must interact with each other over distances of ångströms for electrons to transfer efficiently down the respiratory chain. It's hard to imagine a more preposterous arrangement for a process so central to all living things – respiration, the vital force!

How did matters come to such a pass? There are few better examples of the short-sightedness of evolution. This crazy solution was probably inevitable. Recall the starting point – bacteria that live inside other bacteria. Without such an endosymbiosis, we saw that complex life is not possible, as only autonomous cells are capable of losing superfluous genes, leaving them ultimately with only those genes necessary to control respiration locally. That sounds reasonable enough, but the only limit on gene loss is natural selection – and selection acts on both the host cells and mitochondria. What leads to gene loss? In part, simply replication speed: the bacteria with the smallest genomes replicate the fastest, hence tend to dominate over time. But replication speed cannot explain the transfer of genes to the nucleus, only the loss of genes from mitochondria. In the previous chapter, we saw why mitochondrial genes arrive in the nucleus – some mitochondria die, spilling their DNA into the host cell, and this is taken up into the nucleus. That's hard to stop. Some of this DNA in the nucleus now acquires a targeting sequence, an address code, which targets the protein back to the mitochondria.

This may sound like a freak event, but in fact it applies to almost all the 1,500 known proteins targeted to the mitochondria; plainly it is not so hard. There must be a transient situation where copies of the same gene are present in the surviving mitochondria and in the nucleus at the same time. In the end one of the two copies is lost. Excepting the 13 protein-coding genes that remain in our mitochondria (<1% of their initial genome), the nuclear copy was retained and the mitochondrial copy lost in every case. That doesn't sound like chance. Why is the nuclear copy favoured? There are various plausible reasons, but theoretical work has not yet proved the case one way or another. One possible reason is male fitness. As mitochondria pass down the female line, from mother to daughter, it is not possible to select for mitochondrial variants that favour male fitness, as any genes in male mitochondria that happen to improve male fitness are never passed on. Transferring these mitochondrial genes to the nucleus, where they are passed on in both males and females, could therefore improve male fitness as well as female fitness. Genes in the nucleus are also recombined by sex every generation, perhaps improving fitness even more. And then there's the fact that mitochondrial genes physically take up space, which could be better filled with the machinery of respiration or other processes. Finally, reactive free radicals escape from respiration, which can mutate the neighbouring mitochondrial DNA; we will return to the effects of free radicals on cell physiology later. All in all, there are very good reasons why genes are transferred from the mitochondria to the nucleus; from this point of view, it's more surprising that any genes remain there at all.

Why do they? The balancing force, which we discussed in Chapter 5, is the requirement for genes to control respiration locally. Recall that the electrical potential across the thin inner mitochondrial membrane is 150–200 millivolts, giving a field strength of 30 million volts per metre, equal to a bolt of lightning. Genes are needed to control this colossal membrane potential in response to changes in electron flux, oxygen availability, ADP and ATP ratios, number of respiratory proteins, and more. If a gene that is needed for controlling respiration in this way is transferred to the nucleus, and its protein product fails to make it back to the mitochondria in time to prevent a catastrophe, then the natural 'experiment' ends right there. Animals (and plants) which did not transfer that particular gene to the nucleus survive, while those that transferred the wrong gene die, their unfortunately misconfigured genes with them.

Selection is blind and merciless. Genes are continuously transferred from the mitochondria to the nucleus. Either the new arrangement works better, and the gene stays in its new home, or it does not, and some penalty is exacted – probably death. In the end, nearly all the mitochondrial genes were either lost altogether or transferred to the nucleus, leaving a handful of critical genes in the mitochondria. This is the basis of our mosaic respiratory chains – blind selection. It works. I doubt that an intelligent engineer would have designed it that way; but this was, I hazard, the only way that natural selection could fashion a complex cell, given the requirement for an endosymbiosis between bacteria. This preposterous solution was necessary. In this chapter, we'll examine the consequences of mosaic mitochondria: to what extent does this requirement *predict* the traits of complex cells? I will argue that selection for mosaic mitochondria can indeed explain some of the most puzzling common traits of eukaryotes. All of us. The predicted outcomes of selection include effects on our health, our fitness, fertility and longevity, even our history as a species.

On the origin of species

How and where does selection act? We know that it does. The smoking gun of many gene sequences testifies to a history of selection for coadaptation of mitochondrial and nuclear genes: the two sets of genes change in related ways. We can compare the rates of change of mitochondrial and nuclear genes over time – say, the millions of years that separate chimpanzees from humans or gorillas. We immediately see that the genes which interact directly with each other – those that encode proteins in the respiratory chain, for example – change at about the same speed, whereas other genes in the nucleus generally change (evolve) much more slowly. Plainly a change in a mitochondrial gene tends to elicit a compensatory change in an interacting nuclear gene, or vice versa. So we know that some form of selection has taken place; the question is, what processes fashioned such coadaptation?

The answer lies in the biophysics of the respiratory chain itself. Picture what happens if the nuclear and mitochondrial genomes do *not* match properly. Electrons enter the respiratory chain as normal, but the mismatched genomes encode proteins that do not sit comfortably together. Some electrical interactions between amino acids (hydrogen bonds) are disrupted, meaning that one or two redox centres might now be an ångström further apart than normal. As a result, the electrons flow down the respiratory chain to oxygen at a fraction of their normal speed. They begin to accumulate in the first few redox centres, unable to move on, as the downstream redox centres are already occupied. The respiratory chain becomes highly

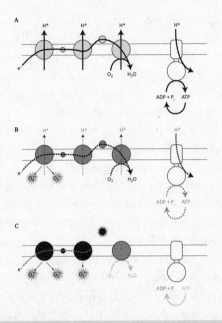

Figure 32 **Mitochondria in cell death**

A shows normal electron flow down the respiratory chain to oxygen (wavy arrow), with the current of electrons powering the extrusion of protons across the membrane, and proton flux through the ATP synthase (right) driving ATP synthesis. The pale grey colour of the three respiratory proteins in the membrane indicates that the complexes are not highly reduced, as electrons do not accumulate in the complexes but are passed on quickly to oxygen. **B** shows the concerted effects of slowing electron flux as the result of an incompatibility between mitochondrial and nuclear genomes. Slow electron flux translates into lower oxygen consumption, limited proton pumping, falling membrane potential (because fewer protons are pumped), and collapsing ATP synthesis. The accumulation of electrons in the respiratory chain is signified by the darker shading of the protein complexes. The highly reduced state of complex I increases its reactivity with oxygen, forming free radicals such as superoxide $(O_2^{\cdot-})$. **C** If this situation is not resolved within minutes, then free radicals react with membrane lipids including cardiolipin, resulting in the release of cytochrome c (the small protein loosely associated with the membrane in **A** and **B** and now released in **C**. Loss of cytochrome c precludes electron flux to oxygen altogether, reducing the respiratory complexes even more (now shown in black), increasing free-radical leak, and collapsing membrane potential and ATP synthesis. These factors together trigger the cell death pathway, resulting in apoptosis.

reduced, meaning that the redox centres fill up with electrons (**Figure 32**). The first few redox centres are iron-sulphur clusters. The iron is converted from the Fe^{3+} to the Fe^{2+} (or reduced) form, which can react directly with oxygen to form the negatively charged superoxide radical $O_2^{\cdot-}$. The dot here symbolises a single unpaired electron, the defining signature of a free radical. And that puts the cat among the pigeons.

There are various mechanisms, notably the enzyme superoxide dismutase, which quickly eliminate an accumulation of superoxide radicals. But the abundance of such enzymes is carefully calibrated. Too much would risk inactivating a vitally important local signal, which works a bit like a fire alarm. Free radicals act like smoke: eliminate the smoke and you don't solve the problem. In this case, the trouble is that the two genomes do not function well together. Electron flow is impaired, and this generates superoxide radicals – a smoke signal.[2] Above some threshold, free radicals oxidise nearby membrane lipids, notably cardiolipin, resulting in release of the respiratory protein cytochrome c, which is normally loosely tethered to cardiolipin. That scuppers the flow of electrons altogether, as they need to hop on to cytochrome c to get to oxygen. Remove cytochrome c and electrons can no longer reach the end of the respiratory chain. With no electron flow there can be no further proton pumping, and that means the electrical

---

2 Most free-radical leak actually derives from complex I. The spacing between the redox centres in complex I suggests that this is deliberate. Recall the principle of quantum tunnelling: electrons 'hop' from one centre to another, with a probability that depends on the distance, the occupancy and the 'tug' of oxygen (the reduction potential). Within complex I there is an early branch in the path of electron flux. In the main pathway, most centres are spaced within about 11 Å of each other, hence electrons will usually hop rapidly from one to the next. The alternative pathway is a cul-de-sac – electrons can enter, but can't easily leave again. At the branch point, electrons have a 'choice': it is about 8 Å to the next redox centre in the main path, and 12 Å to the alternative centre (**Figure 8**). Under normal circumstances, electrons will flux down the main path. But if that pathway clogs up with electrons – if it is highly reduced – the alternative centre now accumulates electrons. This alternative centre is peripheral and easily reacts with oxygen to produce superoxide radicals. Measurements show that this FeS cluster is the main source of free-radical leak from the respiratory chain. I view this as a mechanism to *promote* free-radical leak as a 'smoke signal' if electron flux is too slow to meet demand.

membrane potential will soon collapse. Thus we have three alterations to electron flux in respiration: first, electron transfer slows down, and so the rate of ATP synthesis also falls. Second, the highly reduced iron–sulphur clusters react with oxygen to produce a burst of free radicals, resulting in the release of cytochrome c from its tethering to the membrane. And third, if nothing is done to compensate for these changes, the membrane potential collapses (**Figure 32**).

I have just described a curious set of circumstances first discovered in the mid 1990s and greeted at the time with 'general stupefaction'. This is the trigger for programmed cell death, or apoptosis. When a cell undergoes apoptosis, it kills itself via a carefully choreographed ballet, the cellular equivalent of the dying swan. Far from simply falling to pieces and decomposing, in apoptosis an army of protein executioners, called caspase enzymes, is set loose from within. These cut up the giant molecules of the cell – DNA, RNA, carbohydrates and proteins – into bits and pieces. The pieces are bound up in little packets of membrane, blebs, and fed to surrounding cells. Within a few hours, all traces of its former existence have gone, airbrushed from history as effectively as a KGB cover-up at the Bolshoi.

Apoptosis makes perfect sense in the context of a multicellular organism. It is necessary for sculpting tissues during embryonic development, and removing and replacing damaged cells. What came as a complete surprise was the central involvement of mitochondria, especially the *bona fide* respiratory protein cytochrome c. Why on earth would the loss of cytochrome c from the mitochondria act as a signal for cell death? Since the discovery of this mechanism, the mystery has only deepened. It turns out that this same combination of events – falling ATP levels, free-radical leak, loss of cytochrome c, and a collapse of membrane potential – is conserved right across eukaryotes. Plant cells and yeast kill themselves in response to exactly the same signal. Nobody expected that. Yet it emerges from first principles, as an inevitable consequence of selection for two genomes – it is *predictably* a universal property of complex life.

Let's think back to our electrons making their way down a mismatched respiratory chain. If the mitochondrial and nuclear genes don't work properly together, the natural biophysical outcome is apoptosis. This is a beautiful example of natural selection honing a process that cannot be stopped from happening: a natural tendency is elaborated by selection, ultimately becoming a sophisticated genetic mechanism, which retains at its heart a clue to its origin. Two genomes are needed for large complex cells to exist at all. They must work well together, or respiration will fail. If they don't work properly together, the cell is eliminated by apoptosis. This can now be seen as a form of functional selection against cells with mismatched genomes. Once again, as the Russian-born geneticist Theodosius Dobzhansky famously observed, nothing in biology makes sense except in the light of evolution.

So we have a mechanism for the elimination of cells with mismatched genomes. Conversely, cells with genomes that do work well together will not be eliminated by selection. Over evolution, the outcome is precisely what we see: the coadaptation of mitochondrial and nuclear genomes, such that sequence changes in one genome are compensated for by sequence changes in the other. As noted in the previous chapter, the existence of two sexes increases the variance between female germ cells – different egg cells contain mainly clonal populations of mitochondria, with different eggs amplifying different clones of mitochondria. Some of these clones will happen to work well against the new nuclear background of the fertilised egg, others less well. Those that don't work sufficiently well are eliminated by apoptosis; those that do work well together survive.

Survive what exactly? In multicellular organisms, the broad answer is development. From a fertilised egg cell (zygote), cells divide to form a new individual. The process is exquisitely controlled. Cells that die unexpectedly by apoptosis during development jeopardise the entire developmental programme and may result in miscarriage, a failure of embryonic development. That isn't necessarily a bad thing. Much better, from the dispassionate view of natural selection, to halt development early, before too many resources have been dedicated to the new individual, than to allow development to run to full term. In the latter case, the offspring would be born with incompatibilities between the nuclear and mitochondrial genes, potentially causing a mitochondrial disease, breakdown of health, and an early death.

On the other hand, terminating development early – sacrificing an embryo if it shows serious incompatibilities between mitochondrial and nuclear genomes – obviously reduces fertility. If a high proportion of embryos fail to develop to full term, the outcome is infertility. The costs and benefits here are absolutely central to natural selection: fitness versus fertility. Plainly there must be refined controls over which incompatibilities trigger apoptosis and death, and which ones are tolerated.

All this may seem a little dry and theoretical. Does it actually matter? Yes! – at least in a few cases, and these could be the tip of an iceberg. The best example comes from Ron Burton, at the Scripps Marine Research Institute, who has been working on mitochondrial–nuclear incompatibilities in the marine copepod *Tigriopus californicus* for more than a decade. Copepods are small crustaceans, 1–2 mm in length, found in almost all wet environments; in this case in the intertidal pools of Santa Cruz Island in southern California. Burton has been crossing between two different populations of these copepods from opposite sides of the island, which were reproductively isolated for thousands of years, despite living only a few miles apart. Burton and his colleagues catalogue what is known as 'hybrid breakdown' in matings between the two populations. Intriguingly, there is little effect in the first generation, the result of a single cross between the two populations; but if the female hybrid offspring are then mated with a male from the original paternal population, her own offspring are terribly sickly, in a 'sorry state' to borrow from the title of one of Burton's papers. While there was quite a spectrum of outcomes, on average the hybrid fitness was substantially lower – their ATP synthesis was reduced by about 40%, and this was reflected in similar decreases in survival, fertility and developmental time (in this case, time to metamorphosis, which depends on body size, hence growth rate).

This entire problem could be ascribed to incompatibilities between mitochondrial and nuclear genes by a simple expedient – backcrossing male hybrid offspring with females from the original maternal population. Their offspring were now restored to full and normal fitness. In the opposite experiment, however – crossing female hybrid offspring with males from the original paternal population – there was no positive effect on fitness.

The offspring remained sickly; indeed they were worse than ever. The results are easy enough to understand. The mitochondria always come from the mother, and to function properly need to interact with genes in the nucleus that are similar to those of the mother. By crossing with males from a genetically distinct population, the mother's mitochondria are paired with nuclear genes that don't function well with her mitochondria. In the first generation cross, the problem is not too severe, as 50% of nuclear genes still come from the mother, and these work well with her mitochondria. By the second generation of hybrids, though, 75% of nuclear genes are now mismatched to the mitochondria, and we see a serious breakdown in fitness. Crossing hybrid males with females from the original maternal population means that 62.5% of nuclear genes now derive from the maternal population and *match* the mitochondria. Full health is restored. But the reverse cross has the opposite effect: the maternal mitochondria are now *mismatched* with some 87.5% of nuclear genes. No wonder they were a sickly bunch.

Hybrid breakdown. Most of us are familiar with the idea of hybrid vigour. Outcrossing is beneficial, because unrelated individuals are less likely to share the same mutations in the same genes, so the copies inherited from the father and mother are likely to be complementary, improving fitness. But hybrid vigour only goes so far. Crosses between distinct species are likely to produce offspring that are unviable or sterile. This is hybrid breakdown. The sexual barriers between closely related species are far more permeable than the textbooks would have us believe – species that prefer to ignore each other in the wild, for behavioural reasons, will often mate successfully in captivity. The traditional definition of a species – the failure to produce fertile offspring in crosses between populations – is simply not true for many closely related species. Nonetheless, as populations diverge over time, reproductive barriers do build up between them, and ultimately such crosses do indeed fail to produce fertile offspring. These barriers must begin to assert themselves in crosses between populations of the same species that have been reproductively isolated for long periods, as in Ron Burton's copepods. In this case, the breakdown is entirely attributable to incompatibilities between mitochondrial and nuclear genes. Could similar incompatibilities cause hybrid breakdown in the origin of species more generally?

I suspect so. This is of course only one mechanism among many, but other examples of 'mitonuclear' breakdown have been reported across many species, from flies and wasps to wheat, yeast, and even mice. The fact that this mechanism emerges from a *requirement* for two genomes to work properly together implies that speciation follows inevitably in eukaryotes. Even so, the effects are sometimes more pronounced than others. The reason apparently relates to the rate of change of mitochondrial genes. In the case of copepods, the mitochondrial genes evolve up to 50 times faster than genes in the nucleus. In the case of the fruit fly *Drosophila*, however, the mitochondrial genes evolve much more slowly, barely twice the speed of genes in the nucleus. Accordingly, mitonuclear breakdown is more serious in copepods than in fruit flies. The faster rate of change translates into more differences in sequence within a given time, hence a greater likelihood of incompatibilities between genomes in crosses between different populations.

Exactly why the mitochondrial genes of animals evolve much faster than nuclear genes is unknown. Doug Wallace, the inspirational pioneer of mitochondrial genetics, argues that mitochondria are the front line in adaptation. Rapid changes in mitochondrial genes enable animals to adapt swiftly to changing diets and climates, the first steps that precede slower morphological adaptations. I like the idea, although there is as yet little good evidence either for or against it. But if Wallace is right, then adaptation is improved by continually throwing up new variants in mitochondrial sequence on which selection can act. These changes, in being the first to facilitate adaptations to new environments, are also among the first harbingers of speciation. That corresponds to a curious old rule in biology, first laid down by the inimitable J.B.S. Haldane, one of the founding fathers of evolutionary biology. A new interpretation of this rule suggests that mitonuclear coadaptation might indeed play an important role in the origin of species, and in our own health.

Sex determination and Haldane's rule

Haldane was given to memorable pronouncements, and in 1922 he came up with this remarkable proclamation:

> When in the offspring of two different animal races one sex is absent, rare, or sterile, that sex is the heterozygous [heterogametic] sex.

It would have been easier if he had said the 'male', but that would actually have been less sweeping. The male is heterozygous or heterogametic in mammals, meaning that the male has two different sex chromosomes – an X and a Y chromosome. Female mammals have two X chromosomes, and are therefore homozygous for their sex chromosomes (homogametic). It's the other way round for birds and some insects. Here the female is heterogametic, having a W and a Z chromosome, whereas the male is homogametic, with two Z chromosomes. Imagine a cross between a male and female from two closely related species, which produces viable offspring. But now look more carefully at these offspring: they are all male, or all female; or if both sexes are present, then one of the two sexes is sterile or otherwise maimed. Haldane's rule says that this sex will be the male in mammals and the female in birds. The catalogue of examples that has been pieced together since 1922 is impressive: hundreds of cases conform to the rule, across many phyla, with surprisingly few exceptions for a subject as confounded by exceptions as biology.

There have been various plausible explanations for Haldane's rule, though none of them can account for all cases, hence none of them is intellectually entirely satisfying. For example, sexual selection is stronger in males, which must compete among themselves for the attentions of females (technically there is greater variance in reproductive success between males than females, making male sexual traits more visible to selection). That in turn makes males more vulnerable to hybrid breakdown in a cross between different populations. The trouble is that this particular explanation does not explain why male birds are less vulnerable to hybrid breakdown than females.

Another difficulty is that Haldane's rule arguably goes beyond mere sex chromosomes, which look parochial on a wider view of evolution. Many reptiles and amphibians don't have sex chromosomes at all, but define their sexes on the basis of temperature; eggs incubated at a warmer temperature develop into males, or occasionally vice versa. In fact, given its apparently basic importance, the mechanisms of sex determination are perplexingly variable across species. Sex can be determined by parasites, or by the number of chromosomes, or by hormones, environmental triggers, stress, population density, or even mitochondria. The fact that one of the two sexes tends to be worse affected in crosses between populations, even when sex is not determined by any chromosomes at all, suggests there might be a deeper mechanism afoot. Indeed, the very fact that the detailed mechanisms of sex determination are so variable, yet the development of two sexes is so consistent, implies that there could be some conserved underlying basis to sex determination (the process driving male or female development) which different genes merely embroider.

One possible underlying basis is metabolic rate. Even the ancient Greeks appreciated that males are, literally, hotter than females – the 'hot male' hypothesis. In mammals such as humans and mice, the earliest distinction between the two sexes is the growth rate: male embryos grow slightly faster than females, a difference that can be measured within hours of conception, using a ruler (but definitely don't try that at home). On the Y chromosome, the gene that dictates male development in humans, the *SRY* gene, speeds up the growth rate by turning on a number of growth factors. There is nothing sex specific about these growth factors: they are normally active in both males and females, it's just that their activity is set at a higher level in males than females. Mutations that increase the activity of these growth factors, speeding growth rate, can induce a sex change, forcing male development on female embryos lacking a Y chromosome (or *SRY* gene). Conversely, mutations that lower their activity can have the opposite effect, converting males with a perfectly functional Y chromosome into females. All this implies that growth rate is the real force behind sexual development, at least in mammals. The genes are just holding the reins, and can easily be replaced over evolution – one gene that sets the growth rate is replaced by a different gene that sets the same growth rate.

The idea that males have a faster growth rate corresponds intriguingly to the fact that temperature determines which sex develops in amphibians and reptiles such as alligators. This is related because metabolic rate also depends in part on temperature. Within limits, raising the body temperature of a reptile by 10°C (by basking in the sun, for example) roughly doubles the metabolic rate, which in turn sustains a faster growth rate. While it's not always the case that males develop at the higher temperature (for various subtle reasons) the link between sex and growth rate, set either by genes or temperature, is more deeply conserved than any particular mechanism. It does look as if various opportunistic genes have from time to time grabbed the reins of developmental control, setting a developmental rate that induces male or female development. This is one reason, incidentally, why men need not fear the demise of the Y chromosome – its function will probably be taken over by some other factor, possibly a gene on a different chromosome, which sets the faster metabolic rate needed for male development. It might also account for the strange vulnerability of external testicles in mammals; getting the temperature right is much more deeply embedded in our biology than scrota.

These ideas, I must say, came as a revelation to me. The hypothesis that sex is determined ultimately by metabolic rate has been advanced over several decades by Ursula Mittwoch, a colleague at UCL, who is still remarkably active and publishing important papers at the age of 90. Her papers are not as well known as they should be, perhaps because measurements of 'unsophisticated' parameters such as growth rate, embryo size, and gonadal DNA and protein content seemed old-fashioned in the era of molecular biology and gene sequencing. Now that we are entering a new era of epigenetics (what factors control gene expression?) her ideas resonate better, and I hope will take their proper place in the history of biology.[3]

---

3 Mittwoch points out a parallel problem relating to true hermaphrodites – people who are born with both types of sex organ, for example a testis on the right-hand side and an ovary on the left. It's far more likely to be that way round. Barely a third of people with true hermaphroditism have the testis on the left-hand side and the ovary on the right. The difference can hardly be genetic. Mittwoch shows that at critical periods,

But how does all this relate to Haldane's rule? Sterility or inviability corresponds to a loss of function. Beyond some threshold, the organ or the organism fails. The limits of function depend on two simple criteria – the metabolic demands required to accomplish the task (making sperm, or whatever) and the metabolic power available. If the power available is lower than that required, then the organ or organism dies. In the subtle world of gene networks, these may seem to be absurdly blunt criteria, but they're nonetheless important for that. Put a plastic bag over your head, and you cut off your metabolic power in relation to your needs. Function ceases in little more than a minute, at least in the brain. The metabolic requirements of your brain and heart are high; they will be the first to die. Cells in your skin or intestines might survive much longer; they have much lower requirements. The residual oxygen will be sufficient to meet their low metabolic needs for hours or perhaps even days. From the view point of our constituent cells, death is not all or nothing, it is a continuum. We are a constellation of cells, and they do not all die at once. Those with the highest demands generally fail to meet them first.

This is precisely the problem in mitochondrial diseases. Most involve neuromuscular degeneration, affecting the brain and skeletal muscle, essentially the tissues with the highest metabolic rate. Sight is especially vulnerable: the metabolic rate of cells in the retina and optic nerve is the highest in the body, and mitochondrial diseases such as Leber's hereditary optic neuropathy affect the optic nerve, causing blindness. It is difficult to generalise about mitochondrial diseases, as their severity depends on many factors – the type of mutation, the number of mutants, and also their segregation between tissues. But setting that aside, the fact remains that mitochondrial diseases mostly affect the tissues with the highest metabolic demands.

Imagine that the number and type of mitochondria in two cells are the same, giving them matching capacity to generate ATP. If the metabolic demands imposed on these two cells are different, the outcome will be

---

the right-hand side grows slightly faster than the left, and so is more likely to develop maleness. Curiously, in mice it is exactly the other way around – the left-hand side grows slightly faster and is more likely to develop testes.

different (**Figure 33**). For the first cell, let's say its metabolic demands are low: it meets them comfortably, producing more than enough ATP, and spending it on whatever its task may be. Now imagine that the demands on the second cell are much greater – actually higher than its maximal capacity for producing ATP. The cell strains to match its demands, its whole physiology geared to meet this high output. Electrons pour into the respiratory chains, but their capacity is too low: the electrons are entering faster than they can leave again. The redox centres become highly reduced, and react with oxygen to produce free radicals. These oxidise the surrounding membrane lipids, releasing cytochrome c. Membrane potential falls. The cell dies by apoptosis. This is still a form of functional selection, even in a tissue setting, as cells that can't meet their metabolic demands are eliminated, leaving those that can.

Of course, the removal of cells that don't work well enough only improves overall tissue function if they are replaced with new cells from the stem-cell population. A major problem with neurons and muscle cells is that they cannot be replaced. How could a neuron be replaced? Our life's experience is written into synaptic networks, each neuron forming as many as 10,000 different synapses. If the neuron dies by apoptosis, those synaptic connections are lost forever, along with all the experience and personality that might have been written into them. That neuron is irreplaceable. In fact, while less obviously necessary, all terminally differentiated tissues are irreplaceable – their very existence is impossible without the deep distinction between germline and soma discussed in the previous chapter. Selection is all about offspring. If organisms with large and irreplaceable brains leave more viable offspring than organisms with small replaceable brains, then they will thrive. Only when there is a distinction between the germline and soma can selection act in this way; but when it does, the body becomes disposable. Lifespan becomes finite. And cells that can't meet their metabolic requirements will kill us in the end.

That's why the metabolic rate matters. Cells with a faster metabolic rate are more likely to fail to meet their demands, given the same mitochondrial output. Not only mitochondrial diseases, but also normal ageing and age-related diseases, are most likely to affect the tissues with the

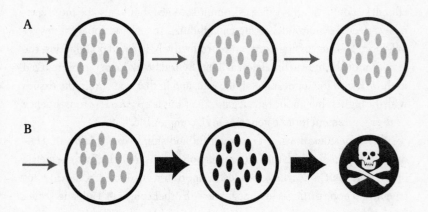

Figure 33 **Fate depends on capacity to meet demand**
Two cells with equivalent mitochondrial capacity, facing different demands. In **A**
the demand is moderate (signified by the arrows); the mitochondria can meet it
comfortably without becoming highly reduced (denoted by the pale grey
shading). In **B** the initial demand is moderate, but then increases to a far higher
level. Electron input to the mitochondria increases commensurately, but their
capacity is insufficient and the respiratory complexes become highly reduced
(dark shading). Unless capacity can be increased swiftly, the outcome is cell death
by apoptosis (as depicted in Figure 32).

highest metabolic demands. And to come full circle – the sex with the highest metabolic demands. Males have a faster metabolic rate than females (in mammals at least). If there is some genetic defect in the mitochondria, that defect will be unmasked largely in the sex with the faster metabolic rate – the male. Some mitochondrial diseases are indeed more common in men than in women; Leber's hereditary optic neuropathy, for example, is five times more prevalent in men, while Parkinson's disease, which also has a strong mitochondrial component, is twice as common. Males should also be more seriously affected by mitonuclear incompatibilities. If such incompatibilities are produced by outcrossing between reproductively isolated populations, the result should be hybrid breakdown. Thus hybrid breakdown is most marked in the sex with the highest metabolic rate, and in that sex, within the tissues with the highest metabolic rate. Again, all of this is a *predictable* consequence of the requirement for two genomes in all complex life.

These considerations offer a beautiful and simple explanation of Haldane's rule: the sex with the fastest metabolic rate is more likely to be sterile or inviable. But is it true, or indeed important? An idea can be true but trivial, and none of this is incompatible with other causes of Haldane's rule. There is nothing to say that metabolic rate should be the sole cause; but is it a significant contributor? I think so. Temperature is well known to exacerbate hybrid breakdown, for example. When the flour beetle *Tribolium castaneum* is crossed with a closely related species *Tribolium freeman*, the hybrid offspring are healthy at their normal rearing temperature of 29°C – but if they are raised at 34°C, the females (in this case) develop deformities in their legs and antenna. That kind of sensitivity to temperature is widespread, often causing sex-specific sterility, and is most easily understood in terms of metabolic rate. Above a certain threshold of demand, particular tissues will start to break down.

Those particular tissues often include the sex organs, especially in males, where sperm production continues throughout life. A striking example is found in plants, known as cytoplasmic male sterility. Most flowering plants are hermaphrodites, but a large proportion exhibit male sterility, giving them two 'sexes' – hermaphrodites and (male-sterilized) females. This mishap is caused by mitochondria and has usually been interpreted in terms of selfish conflict.[4] But molecular data suggest that male sterility may simply reflect metabolic rate. The plant scientist Chris Leaver, in Oxford, has shown that the cause of cytoplasmic male sterility in sunflowers is a gene encoding a single subunit of the ATP synthase enzyme in the mitochondria. The problem in this case is an error in recombination, which affects a relatively small proportion (importantly, not all) of the ATP synthase enzymes. That lowers the maximum rate of ATP synthesis. In most tissues, the effects of this mutation are unnoticeable – only the male sex organs, the anthers, actually degenerate. They degenerate because their constituent cells die by apoptosis, involving the release of cytochrome *c* from their mitochondria in exactly the same way as in ourselves. The anthers seem to be the only tissue in the sunflower with a metabolic rate that is high enough to trigger degeneration: only there do the faulty mitochondria fail to meet their metabolic demands. The outcome is male-specific sterility.

Similar findings have been reported in the fruit fly *Drosophila*. By transferring the nucleus from one cell to another, it's possible to construct hybrid cells (cybrids) in which the nuclear genome is more or less identical, but the mitochondrial genes differ.[5] Doing this with egg cells produces embryonic

---

4 The mitochondria pass down the female line, in the egg cells, not sperm. Hermaphrodites are theoretically particularly vulnerable to sex distortion by mitochondria. From their point of view, the male is a genetic dead end – the last place the mitochondria 'want' to end up is in the anthers. It is therefore in their interests to sterilise the male sex organs, to ensure their passage in a female plant. Many bacterial parasites in insects, notably *Buchnera* and *Wolbachia*, play a similar game – they can completely distort sex ratios in insects by selectively killing males. The central importance of mitochondria to the host organism means that they have less scope than bacterial parasites to kill males through such selfish conflict, but they might nonetheless cause sterility or selective damage to males. However, I'm inclined to think conflict plays a lesser role in Haldane's rule, as it cannot explain why females should be worse affected in birds (and flour beetles).

5 Such cybrids are widely used in cell culture experiments, as they allow precise measurements to be made of cell function, notably respiration. Mismatching mitochondrial and nuclear genes between species reduces the rate of respiration, and as noted, increases free-radical leak. The magnitude of functional deficit depends on genetic distance. Cybrids constructed from chimp mitochondrial DNA and human nuclear genes

flies that are genetically identical in their nuclear background, but which have mitochondrial genes from related species. The outcomes are strikingly different, depending on the mitochondrial genes. In the best cases, there is nothing wrong with the newborn flies. In the worst crosses, the males are sterile, the male being the heterogametic sex in *Drosophila*. Most interesting are the intermediate cases, where the flies seem to be fine. A closer look at the activity of genes in various organs shows that they are troubled in their testes, however. More than 1,000 genes in the testes and accessory sexual organs are up-regulated in male flies. Quite what is going on is not well understood, but the simplest explanation, in my eyes, is that these organs can't really cope with the metabolic demands placed upon them. Their mitochondria are not wholly compatible with genes in the nucleus. Cells in the testes, with their high metabolic demands, are physiologically stressed, and this stress produces a response that involves a substantial part of the genome. As in cytoplasmic male sterility in plants, only the metabolically challenged sex organs are affected – and only in males.[6]

If all that's the case, why are females affected in birds? Roughly the same reasoning holds, but with some intriguing differences. In a few birds, notably birds of prey, the female is larger than the male, and so presumably grows faster. But that is not universal. Ursula Mittwoch's early work shows that in chickens the ovaries outgrow the testes, after a slow start for the first week or so. In these cases the prediction would be that female birds should suffer from sterility rather than inviability, as only their sex organs grow faster. But that is not true. Most cases of Haldane's rule in birds in fact seem to be inviability rather than sterility. I was baffled by that until last year

---

(yes, it has been done, but only in cell culture) show that the rate of ATP synthesis is about half that of normal cells. Cybrids between mice and rats don't have functional respiration at all.

6 This conjecture might seem a little odd: do the testes really have a higher metabolic rate than other tissues such as the heart, brain or flight muscle? Not necessarily. The problem is the capacity to meet demand. It might be that peak demand is indeed greater in testes, or that the number of mitochondria that are called upon to meet that demand is lower, so that the demand per mitochondrion is greater. This is a simple testable prediction, but to my knowledge it has not been tested.

when Geoff Hill, a specialist in sexual selection in birds, sent me his paper on Haldane's rule in birds. Hill pointed out that a few nuclear genes encoding respiratory proteins in birds are found on the Z chromosome (recall that, in birds, males have two Z chromosomes, while females have one Z and one W chromosome, making them the heterogametic sex). Why does that matter? If female birds only inherit one copy of the Z chromosome, they only get one copy of several critical mitochondrial genes and they inherit them from their father. If the mother did not choose the father with care, then her mitochondrial genes might not match the single copy of his nuclear genes. Breakdown could be immediate and serious.

Hill argues that this arrangement places the burden on the female to select her mate with extreme care, or pay the grave penalty (her female offspring die). That in turn could account for the vibrant plumage and coloration of male birds. If Hill is right, the detailed pattern of the plumage signals the mitochondrial type: sharp demarcations in pattern are postulated to reflect sharp demarcations in mitochondrial DNA type. The female therefore uses pattern as a guide to compatibility. But a male of the right type can still be a pretty poor specimen. Hill argues that the vibrancy of colour reflects mitochondrial function, as most pigments are synthesised in the mitochondria. A brightly coloured male must have top-quality mitochondrial genes. There's little evidence to back this hypothesis at present, but it gives a sense of how pervasive the requirement for mitonuclear coadaptation could turn out to be. It's a sobering thought that the requirement for two genomes in complex life could explain evolutionary conundrums as disparate as the origin of species, the development of sexes, and the vivid coloration of male birds.

And it may run even deeper. There are penalties for getting mitonuclear incompatibility wrong, but also costs to getting it right, achieving good compatibility. The balance of costs and benefits should differ between different species, depending on their aerobic requirements. The trade-off, we'll see, is between fitness and fertility.

The threshold of death

Imagine you can fly. Gram per gram, you have more than twice the power of a cheetah in full flight, a remarkable combination of strength, aerobic capacity, and lightness. You have no hope of getting airborne if your mitochondria are not practically perfect. Consider the competition for space in your flight muscles. You need myofibrils, of course, the sliding filaments that produce muscle contraction. The more of these you can pack in, the stronger you will be, as the strength of a muscle depends on its cross-sectional area, like a rope. Unlike a rope, however, muscle contraction has to be powered by ATP. To sustain exertion for much more than a minute requires ATP synthesis on the spot. That means you need mitochondria right there in your muscle. They take up space that could otherwise be occupied by more myofibrils. Mitochondria also need oxygen. That means capillaries, to deliver oxygen and remove waste. The optimal space distribution in aerobic muscle is around a third for myofibrils, a third for mitochondria, and a third for capillaries. That's true for us, and cheetahs, and humming birds, which have by far the fastest metabolic rates of all vertebrates. The bottom line is we can't get more power just by accumulating more mitochondria.

All this means that the only way birds can generate enough power to remain airborne for long is to have 'supercharged' mitochondria, able to generate more ATP per second per unit of surface area than 'normal' mitochondria. The electron flux from food to oxygen must be fast. That translates into fast proton pumping and a fast rate of ATP synthesis, necessary to sustain the high metabolic rate. Selection must act at every step, speeding up the maximum rate at which each respiratory protein operates. We can measure these rates, and we know that the enzymes in the mitochondria of birds do indeed operate faster than those in mammals. But, as we've seen, the respiratory proteins are mosaics, composed of subunits encoded by two different genomes. Fast electron flux entails strong selection for the two genomes to work well together, for mitonuclear coadaptation. The greater the aerobic requirements, the stronger this selection for coadaptation must be. Cells with genomes that don't work well together are eliminated by apoptosis. The most reasonable place for such selection to occur, as we've seen, is during embryonic development. From a dispassionately theoretical point of view, it makes more sense to terminate embryonic development very early on if the embryo has incompatible genomes, which don't work well enough together to sustain flight.

But how incompatible is incompatible, how bad is bad? Presumably, there must be some sort of a threshold, a point at which apoptosis is triggered. Above that threshold, the speed of electron flux through the mosaic respiratory chain is just not good enough – it's not up to the job. Individual cells, and by extension the whole embryo, die by apoptosis. Conversely, below the threshold, electron flux is fast enough. If so, then it follows that the two genomes must function well together. The cells, and by extension the whole embryo, do not kill themselves. Instead, development continues, and all being well a healthy chick is born, its mitochondria 'pretested' and stamped fit for purpose.[7] The crucial point is that 'fit for purpose' must vary with the purpose. If the purpose is flight, then the genomes have to match nearly perfectly. The cost of high aerobic capacity is low fertility. More embryos that could have survived some lesser purpose must be sacrificed on the altar of perfection. We can even see the consequences in mitochondrial gene sequences. Their rate of change in birds is lower than in most mammals (except bats, which face the same problem as birds). Flightless birds, which don't face the same constraints, have a faster rate of change. The reason that most birds have low rates of change is that they have already perfected their mitochondrial sequence for flight. Changes from this ideal sequence are not readily tolerated, so are typically eliminated by selection. If most change is eliminated, then what remains is relatively unchanging.

What if we embrace a lesser purpose? Let's say I am a rat (as my son's school song goes, 'there's no escaping that') and I have no interest in flying.

---

7 I suspect that the free-radical signal is deliberately amplified at some point during embryonic development. For example, the gas nitric oxide (NO) can bind to cytochrome oxidase, the final complex of the respiratory chain, increasing free-radical leak and the likelihood of apoptosis. If NO were produced in larger amounts at some point during development, the effect would be to amplify the signal above a threshold, terminating embryos with incompatible genomes – a checkpoint.

It would be stupid to sacrifice most of my prospective offspring on the altar of perfection. We've seen that the trigger for apoptosis – functional selection – is free-radical leak. A sluggish flux of electrons in respiration betokens a poor compatibility between the mitochondrial and nuclear genomes. The respiratory chains become highly reduced and leak free radicals. Cytochrome c is released and membrane potential falls. If I were a bird, this combination is the trigger for apoptosis. My offspring would die in embryo again and again. But I'm a rat and I don't want that. What if, by some biochemical sleight of hand, I 'ignore' the free-radical signal that heralds the death of my offspring? I raise the death threshold, meaning I can tolerate more free-radical leak before triggering apoptosis. I gain an immeasurable benefit: most of my offspring survive embryonic development. I become more fertile. What price will I pay for my burgeoning fertility?

Certainly I'm never going to fly. And more generally, my aerobic capacity will be limited. The chance of my offspring having an optimal match between mitochondrial and nuclear genes is remote. That leads straight to another important pairing of costs and benefits – adaptability versus disease. Recall Doug Wallace's hypothesis that the rapid evolution of mitochondrial genes in animals facilitates their adaptation to different diets and climates. We don't really know how, or indeed if, this really works, but it would be surprising if there were no truth in it at all. The first line of adaptation relates to diet and body temperature (we won't survive for long if we can't get the basics right) and mitochondria are absolutely central to both. The performance of mitochondria depends in large measure on their DNA. Different sequences of DNA support different levels of performance. Some will work better in cooler environments than hotter ones, or in greater humidity, or burning up a fatty diet, and so on.

There have been hints from the apparently non-random geographical distribution of different types of mitochondrial DNA in human populations that selection in particular environments might indeed exist, but little more than hints. Yet there is undoubtedly less variation in the mitochondrial DNA of birds, as we've just noted. The very fact that most changes from the optimal sequence for flight are eliminated by selection means that less varied mitochondrial DNA remains – so there is less scope for selection to choose some mitochondrial variant that just happens to be particularly good in the cold, or with a fatty diet. It's curious in this regard that birds frequently migrate rather than suffer a seasonal change in environmental conditions. Could it be that their mitochondria are better able to support the exertion of migration than function in the harsher environment they would face if they stayed put? Conversely, rats have a great deal more variation, and from first principles that should give them the raw material for better adaptation. Does it really? Frankly, I don't know; though rats are pretty adaptable beasts. There's no escaping that.

But, of course, mitochondrial variation comes at a cost – disease. To a point, that can be avoided by selection on the germline, in which egg cells with mitochondrial mutations are weeded out before they can mature. There is some evidence for such selection – severe mitochondrial mutations tend to be eliminated over several generations, though less severe mutations persist almost indefinitely in mice and rats. But think about that remark again – several generations! Selection here is pretty weak. If you are born with a serious mitochondrial disease it can be little consolation to think that your grandchildren, if you are lucky enough to have any, may be disease free. Even if selection does act against mitochondrial mutations in the germline, this is still no guarantee against mitochondrial disease. Immature egg cells do not have an established nuclear background. Not only are they held in limbo for many years, paused halfway through meiosis, but at this point the father's genes have yet to be added to the fray. Selection for mitonuclear coadaptation can only occur after the mature egg cell has been fertilised by the sperm, and a new, genetically unique nucleus has been fashioned. Hybrid breakdown is not caused by mitochondrial mutations, but by incompatibilities between nuclear and mitochondrial genes, all of which are perfectly functional in some other context. We've already seen that strong selection against mitonuclear incompatibilities necessarily increases the likelihood of infertility. If we don't want to be infertile, we have to accept the cost – a greater risk of disease. Again, this equation between fertility and disease is a *predictable* outcome of the requirement for two genomes.

So there is a hypothetical death threshold (**Figure 34**). Above the threshold the cell, and by extension the whole organism, dies by apoptosis. Below

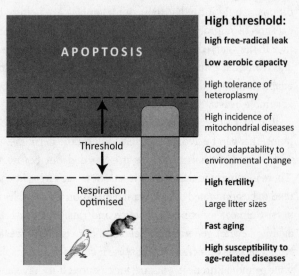

Figure 34 **The death threshold**
The threshold at which free-radical leak triggers cell death (apoptosis) should vary between species, depending on the aerobic capacity. Organisms with high aerobic demands need a very good match between their mitochondrial and nuclear genomes. A poor match is betrayed by a high rate of free-radical leak from the dysfunctional respiratory chain (see Figure 32). If a very good match is required, cells should be more sensitive to free-radical leak; even low leak signals that the match is not good enough, triggering cell death (a low threshold). Conversely, if aerobic demands are low then there is nothing to be gained by killing the cell. Such organisms will tolerate higher levels of free-radical leak without triggering apoptosis (a high threshold). The predictions for high and low death threshold are shown in the side panels. Pigeons are hypothesised to have a low death threshold, rats the opposite. Both have the same body size and basal metabolic rate, but pigeons have a much lower rate of free-radical leak. While the veracity of these predictions is unknown, it is striking rats live for just three or four years, pigeons for up to 30.

the threshold the cell and the organism survive. This threshold is necessarily variable between different species. For bats and birds and other creatures with high aerobic requirements, the threshold must be set low – even a modest rate of free-radical leak from mildly dysfunctional mitochondria (with slight incompatibilities between mitochondrial and nuclear genomes) signals apoptosis and termination of the embryo. For rats and sloths and couch potatoes with low aerobic requirements, the threshold is set higher: a modest rate of free-radical leak is now tolerated, dysfunctional mitochondria are good enough, the embryo develops. There are costs and benefits to both sides. A low threshold gives a high aerobic fitness and a low risk of disease, but at the cost of a high rate of infertility and poor adaptability. A high threshold gives a low aerobic capacity and higher risk of disease but with the benefits of greater fertility and better adaptability. These are words to conjure with. Fertility. Adaptability. Aerobic fitness. Disease. We can't cut much closer to the grain of natural selection than that. I reiterate: all these trade-offs emerge inexorably from the requirement for two genomes.

I just called this a hypothetical death threshold, and so it is. Does it really exist? If so, is it genuinely important? Just think about ourselves. Apparently 40% of pregnancies end in what is known as 'early occult miscarriage'. 'Early' in this context means very early – within the first weeks of pregnancy, and typically before the first overt signs of pregnancy. You'd never know you were pregnant. And 'occult' means 'hidden' – not clinically recognised. Generally we don't know why it happened. It's not caused by any of the usual suspects – chromosomes that failed to separate, giving a 'trisomy' and suchlike. Could the problem be bioenergetic? It is hard to prove that one way or the other, but in this brave new world of fast genome sequencing, it should be possible to find out. The emotional distress of infertility has sanctioned some rather insalubrious research into the factors that promote embryo growth. The shockingly clumsy expedient of injecting ATP into a faltering embryo can prolong its survival. Plainly bioenergetic factors matter. By the same token, perhaps these failures are 'for the best'. Perhaps they had mitonuclear incompatibilities that triggered apoptosis. It is best not to make any moral judgements from evolution. I can only say that I will not forget my own years of shared anguish (thankfully now over), and I, like most people, want to know why. I suspect that a large number of early occult miscarriages do reflect mitonuclear incompatibilities.

But there's another reason to think the death threshold is real and important. There is one final, indirect cost to having a high death threshold – a faster rate of ageing and a greater propensity to suffer age-related diseases. That statement will raise hackles in some quarters. A high threshold means a high tolerance for free-radical leak before triggering apoptosis. That means species with a low aerobic capacity, like rats, should leak more free radicals. And conversely, species with a high aerobic capacity, such as pigeons, should leak fewer free radicals. I choose these species with care. They have almost equal body masses and basal metabolic rates. On that basis alone, most biologists would predict that they should have a similar lifespan. Yet according to the exquisite work of Gustavo Barja in Madrid, pigeons leak fewer free radicals from their mitochondria than do rats.[8] The free-radical theory of ageing argues that ageing is caused by free-radical leak: the faster the rate of free-radical leak, the faster we age. The theory has had a bad decade, but in this case makes a clear prediction – pigeons should live much longer than rats. They do. A rat lives three or four years, a pigeon for nearly three decades. A pigeon is assuredly not a flying rat. So is the free-radical theory of ageing correct? In its original formulation, the answer is easy: no. But I still think a more subtle form is true.

The free-radical theory of ageing

The free-radical theory has its roots in radiation biology in the 1950s. Ionising radiation splits water to produce reactive 'fragments' with single

---

8 Gustavo Barja has found that the rate of free-radical leak is up to 10-fold lower in birds such as pigeons and budgerigars than in rats and mice as a proportion of oxygen consumed. The actual rates vary between tissues. Barja also found that the lipid membranes of birds are more resistant to oxidative damage than those of flightless mammals, and this resistance is reflected in less oxidative damage to DNA and proteins. Altogether, it's hard to interpret Barja's work in any other terms.

unpaired electrons: oxygen free radicals. Some of these, like the notorious hydroxyl radical (OH$^{\cdot}$), are very reactive indeed; others, like the superoxide radical (O$_2^{\cdot-}$) are tame in comparison. The pioneers of free-radical biology – Rebeca Gerschman, Denham Harman and others – realised that the same free radicals can be formed from oxygen directly, deep down in the mitochondria, with no need for radiation at all. They saw free radicals as fundamentally destructive, capable of damaging proteins and mutating DNA. All that is true – they can. Worse than that, they can initiate long chain reactions, in which one molecule after another (typically membrane lipids) grabs an electron, wreaking havoc right across the delicate structures of a cell. Free radicals, the theory went, ultimately cause a crescendo of damage. Picture it. The mitochondria leak free radicals, which react with all kinds of neighbouring molecules including nearby mitochondrial DNA. Mutations accumulate in the mitochondrial DNA, some of which undermine its function, producing respiratory proteins that leak even more radicals. They damage more proteins and DNA, and before long the rot spreads to the nucleus culminating in an 'error catastrophe'. Look at a demographic chart of disease and mortality, and you'll see that their incidence increases exponentially in the decades between 60 and 100. The idea of an error catastrophe (damage feeding on itself) seems to match that graph. And the idea that the whole process of ageing is driven by oxygen, the very gas we need to live, holds the horrifying fascination of a beautiful killer.

If free radicals are bad, antioxidants are good. Antioxidants interfere with the noxious effects of free radicals, blocking the chain reactions, and so preventing the spread of damage. If free radicals cause ageing, antioxidants should slow it down, delaying the onset of diseases and perhaps prolonging our lives. Some celebrated scientists, notably Linus Pauling, bought into the myth of antioxidants, taking several spoonfuls of vitamin C every day. He did live to the ripe old age of 92, but that's still squarely in the normal range, including some people who drank and smoked throughout their lives. Plainly it's not as simple as that.

This black-and-white view of free radicals and antioxidants is still current in many glossy magazines and healthfood stores, even though most researchers in the field realised it was wrong long ago. A favourite quote of mine is from Barry Halliwell and John Gutteridge, authors of the classic textbook *Free Radicals in Biology and Medicine*: 'By the 1990s it was clear that antioxidants are not a panacea for ageing and disease, and only fringe medicine still peddles this notion.'

The free-radical theory of ageing is one of those beautiful ideas killed by ugly facts. And boy, are the facts ugly. Not one tenet of the theory, as it was originally formulated, has withstood the scrutiny of experimental testing. There are no systematic measurements of an increase in free-radical leak from the mitochondria as we age. There is a small increase in the number of mitochondrial mutations, but with the exception of limited regions of tissue, they are typically found at surprisingly low levels, well below those known to cause mitochondrial diseases. Some tissues show evidence of accumulating damage, but nothing that resembles an error catastrophe, and the chain of causality is questionable. Antioxidants most certainly do not prolong life or prevent disease. Quite the contrary. The idea has been so pervasive that hundreds of thousands of patients have enrolled in clinical trials over the past few decades. The findings are clear. Taking high-dose antioxidant supplements carries a modest but consistent risk. You are more likely to die early if you take antioxidant supplements. Many long-lived animals have low levels of antioxidant enzymes in their tissues, while short-lived animals have much higher levels. Bizarrely, *pro*-oxidants can actually extend the lifespan of animals. Taken together, it's not surprising that most of the field of gerontology has moved on. I discussed all this at length in my earlier books. I'd like to think I was prescient in dismissing the notion that antioxidants slow ageing as long ago as 2002, in *Oxygen*, but frankly I wasn't. The writing was on the wall even then. The myth has been perpetrated by a combination of wishful thinking, avarice, and a lack of alternatives.

So why, you may well wonder, do I still think that a more subtle version of the free-radical theory is true? For several reasons. Two critical factors were missing from the original theory: signalling and apoptosis. Free-radical signals are central to cell physiology, including apoptosis, as we've already noted. Blocking free-radical signals with antioxidants is hazardous and can *suppress* ATP synthesis in cell culture, as shown by Antonio Enriques and his colleagues in Madrid. It seems most likely that free-radical signals optimise respiration, within individual mitochondria, by raising the number of respiratory complexes, so increasing respiratory capacity. Because mitochondria spend much of their time fusing together and then splitting apart again, making more complexes (and more copies of mitochondrial DNA) translates into making more mitochondria – what's known as mitochondrial biogenesis.[9] Free-radical leak can therefore increase the number of mitochondria, which between them make more ATP! Conversely, blocking free radicals with antioxidants prevents mitochondrial biogenesis, hence ATP synthesis falls as shown by Enriquez (**Figure 35**). Antioxidants can *undermine* energy availability.

But we've seen that higher rates of free-radical leak, above the death threshold, trigger apoptosis. So are free radicals optimising respiration or eliminating cells by apoptosis? Actually, that's not as contradictory as it sounds. Free radicals signal the problem that respiratory capacity is low, relative to demand. If the problem can be fixed by making more respiratory complexes, raising respiratory capacity, then all is well and good. If that does not fix the problem, the cell kills itself, removing its presumably defective DNA from the mix. If the damaged cell is replaced by a nice new cell (from a stem cell), then the problem has been fixed, or rather, eradicated.

This central role of free-radical signalling in optimising respiration

---

9 I call this 'reactive biogenesis' – individual mitochondria react to a local free-radical signal, which indicates that respiratory capacity is too low to meet demand. The respiratory chain becomes highly reduced (clogged up with electrons). Electrons can escape to react directly with oxygen to produce superoxide radicals. These interact with proteins in the mitochondria that control that the replication and copying of mitochondrial genes, called transcription factors. Some transcription factors are 'redox sensitive', meaning that they contain amino acids (such as cysteine) that can lose or gain electrons, becoming oxidised or reduced. A good example is mitochondrial topoisomerase-1, which controls access of proteins to mitochondrial DNA. The oxidation of a critical cysteine on this protein increases mitochondrial biogenesis. Thus a local free-radical signal (that never leaves the mitochondria) increases mitochondrial capacity, raising ATP production in relation to demand. This kind of local signal in response to sudden changes in demand may explain why mitochondria have retained a small genome (see Chapter 5).

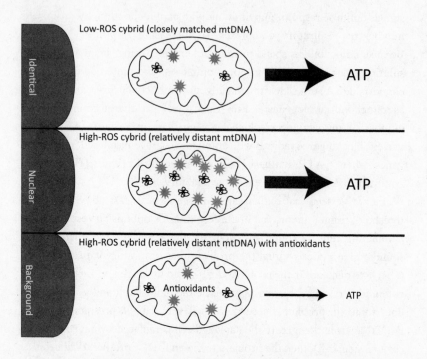

Figure 35 **Antioxidants can be dangerous**

Cartoon depicting the results of an experiment using hybrid cells, or cybrids. In each case the genes in the nucleus are nearly identical; the main difference lies in the mitochondrial DNA. There are two types of mitochondrial DNA: one from the same strain of mice as the nuclear genes (top, 'low ROS'), and the other from a related strain with a number of differences in its mitochondrial DNA (middle, 'high ROS'). ROS stands for reactive oxygen species and equates to the rate of free-radical leak from the mitochondria. The rate of ATP synthesis is depicted by the large arrows, and is equivalent in the low ROS and high ROS cybrids. However, the low ROS cybrid generates this ATP comfortably, with low free-radical leak (denoted by little 'explosions' in the mitochondria) and a low copy number of mitochondrial DNA (squiggles). In contrast, the high-ROS cybrid has more than double the rate of free-radical leak, and double the copy number of mitochondrial DNA. Free-radical leak appears to power-up respiration. That interpretation is supported by the bottom panel: antioxidants lower the rate of free-radical leak, but also reduce the copy number of mitochondrial DNA and, critically, the rate of ATP synthesis. So antioxidants disrupt the free-radical signal that optimises respiration.

explains why antioxidants do not prolong life. They can suppress respiration in cell culture because the normal safeguards imposed by the body are not present in cell culture. In the body, massive doses of antioxidants such as vitamin C are barely absorbed; they tend to cause diarrhoea. Any excess that does get into the blood is swiftly excreted in the urine. Blood levels are stable. That's not to say you should avoid dietary antioxidants, especially vegetables and fruit – you need them. You might even benefit from taking supplementary antioxidants, if you have a poor diet or a vitamin deficiency. But cramming antioxidant supplements on top of a balanced diet (which includes pro-oxidants as well as antioxidants) is counterproductive. If the body allowed high levels of antioxidants into cells, they would cause havoc and potentially kill us through an energy deficiency. So the body doesn't let them in. Their levels are carefully regulated both inside and outside cells.

And apoptosis, by eradicating damaged cells, eliminates the evidence of damage. The combination of free-radical signals and apoptosis together confound most of the predictions of the original free-radical theory of ageing, which was formulated long before either process was known. We don't see a sustained increase in free-radical leak, or a large number of mitochondrial mutations, or an accumulation of oxidative damage, or any benefit to antioxidants, or an error catastrophe, for these reasons. It all makes perfectly reasonable sense, explaining why the predictions of the original free-radical theory of ageing are mostly wrong. But it doesn't give any indication of why the free-radical theory might still be right. Why, if they are so well regulated and beneficial, should free radicals have anything to do with ageing at all?

Well, they *can* explain the variance in lifespan between species. We've known since the 1920s that lifespan tends to vary with metabolic rate. The eccentric biometrician Raymond Pearl entitled an early article on the subject, 'Why lazy people live longer'. They don't; quite the opposite. But this was Pearl's introduction to his famous 'rate-of-living theory', which does have some basis of truth. Animals with a low metabolic rate (often large species such as elephants) typically live longer than animals with a high metabolic rate, like mice and rats.[10] The rule usually holds within major groups such as reptiles, mammals and birds, but doesn't hold at all well between these groups; hence the idea has been somewhat discredited, or at least ignored. But in fact there is a simple explanation, which we've already noted – free-radical leak.

As originally conceived, the free-radical theory of ageing envisioned free radicals as an unavoidable by-product of respiration; about 1–5% of oxygen was thought to be converted unavoidably into free radicals. But that's wrong on two counts. First, all the classical measurements were made on cells or tissues exposed to atmospheric levels of oxygen, much greater than anything that cells are exposed to within the body. The actual rates of leak may be orders of magnitude lower. We just don't know how big a difference this makes in terms of meaningful outcomes. And second, free-radical leak is *not* an unavoidable by-product of respiration – it is a deliberate signal, and the leak rate varies enormously between species, tissues, times of the day, hormonal status, calorie intake, and exercise. When you exercise, you consume more oxygen, so your free-radical leak rises, right? Wrong. It remains similar or even goes down, because the proportion of radicals leaked relative to oxygen consumed falls off considerably. That happens because electron flux speeds up in the respiratory chains, meaning that the respiratory complexes become less reduced, and so are less likely to react directly with oxygen (**Figure 36**). The details don't matter here. The point is that there is no simple relationship between the rate of living and free-radical leak. We've noted that birds live far longer than they really

---

10 This looks like a contradiction – larger species typically have a lower metabolic rate, gram per gram, yet I have talked about male mammals being larger and having a higher metabolic rate, the opposite. Within a species the differences in mass are trivial compared with the many orders of magnitude plotted out between species; on that scale the metabolic rates of adults in the same species are practically the same (though children do have a higher metabolic rate than adults). The sexual differences in metabolic rate that I was talking about earlier relate to differences in absolute growth rates at particular stages of development. If Ursula Mittwoch is correct, these differences are so subtle that they can account for developmental differences on the right versus the left side of the body; see footnote 3.

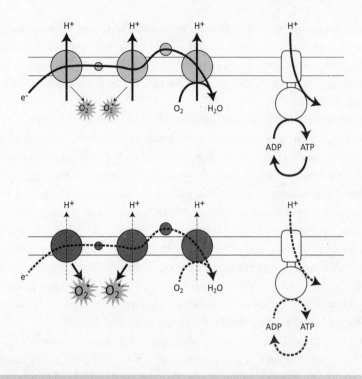

Figure 36 **Why rest is bad for you**

The traditional view of the free-radical theory of ageing is that a small proportion of electrons 'leak' out from the respiratory chain during respiration to react directly with oxygen and form free radicals such as the superoxide radical ($O_2^{\cdot-}$). Because electrons flow faster and we consume more oxygen in active exercise, the assumption has been that free-radical leak increases during exercise, even if the proportion of electrons leaking out remains constant. That is not so. The upper panel here indicates the actual situation during exercise: electron flow down the respiratory chain is fast because ATP is consumed quickly. That allows protons to flux through the ATP synthase, which lowers membrane potential, which allows the respiratory chain to pump more protons, which draws electrons faster down the respiratory chain to oxygen, which prevents the accumulation of electrons in respiratory complexes, lowering their reduction state (represented by the pale grey shading). That means free-radical leak is modest during exercise. The opposite is true at rest (lower panel), meaning there can be higher rates of free-radical leak during *inactivity*. Low consumption of ATP means there is a high membrane potential, it becomes hard to pump protons, so the respiratory complexes gradually fill up with electrons (darker grey shading) and leak more free radicals. Best go for a run.

'ought to' on the basis of their metabolic rate. They have a fast metabolic rate, but leak relatively few free radicals, and live a long time. The underlying correlation is between free-radical leak and lifespan. Correlations are notorious as a guide to causal relationship, but this is an impressive one. Could it be causal?

Consider the consequences of free-radical signalling in the mitochondria: optimising respiration and eliminating dysfunctional mitochondria. The mitochondria that leak the most free radicals will make the most copies of themselves, precisely because free-radical signals correct the respiratory deficit by increasing capacity. But what if the respiratory deficit did not reflect a shift in supply and demand, but rather an incompatibility with the nucleus? Some mitochondrial mutations do occur with ageing, giving rise to a mixture of different mitochondrial types, some of which work better than others with the genes in the nucleus. Think about the problem here. The most *incompatible* mitochondria will tend to leak the most free radicals, and so make more copies of themselves. That can have one of two effects. Either the cell dies by apoptosis, removing its load of mitochondrial mutations, or it doesn't. Let's consider what happens if the cell dies first. Either it's replaced, or it isn't. If it is replaced, all is well. But if it's not replaced, for example in brain or heart muscle, then the tissue will slowly lose mass. Fewer cells remain to do the same job, so they are placed under greater pressure. They become physiologically stressed, with the activity of thousands of genes changing, as in the testes of those fruit flies with mitonuclear incompatibilities. At no stage in this process did free-radical leak necessarily damage proteins or cause an error catastrophe. Everything is driven by subtle free-radical signals within the mitochondria, but the outcome is tissue loss, physiological stress, and changes in gene regulation – all changes associated with ageing.

What happens when the cell doesn't die by apoptosis? If its energy needs are low, they can be met by deficient mitochondria or fermentation to produce lactic acid (what is often, erroneously, called anaerobic respiration). Here we may see the accumulation of mitochondrial mutations in cells that become 'senescent'. These no longer grow, but can be an angry presence in tissues, being stressed themselves, and often provoke chronic inflammation and the dysregulation of growth factors. That stimulates cells that like to grow anyway, stem cells, vascular cells, and so on, goading them to grow when they had better not. If you're unlucky they will develop into cancer, an age-related disease in most cases.

It's worth emphasising again that this whole process is driven by energetic deficiencies that ultimately derive from free-radical signals within the mitochondria. Incompatibilities that accumulate with age undermine mitochondrial performance. This is totally different from the conventional free-radical theory, as it does not call upon oxidative damage in the mitochondria or anywhere else (though of course that doesn't rule it out; it is simply not necessary). As we've noted, because free radicals act as signals that increase ATP synthesis, the *prediction* is that antioxidants should not work: they will not prolong lifespan, nor protect against disease, because they would undermine energy availability if they did ever gain access to mitochondria. This view can also account for the exponential increase in disease and mortality with age: tissue function may decline gently over many decades, eventually falling below the threshold required for normal function. We become less and less able to cope with exertions, and ultimately not even passive existence. This process is recapitulated in everyone, over our dying decades, giving rise to the exponential decline on morticians' graphs.

So what can we do about ageing? I said that Raymond Pearl was wrong: lazy people don't live longer – exercise is beneficial. So, within limits, is calorie restriction and low-carbohydrate diet. All promote a physiological stress response (as do pro-oxidants) which tends to clear out defective cells and bad mitochondria, promoting survival in the short term, but typically at the cost of lowering fertility.[11] Again we see the link between aerobic

---

11 And worse. The best way to clear out bad mitochondria is to force the body to use them, increasing their turnover rate. For example, a high-fat diet tends to force the use of mitochondria, whereas a high-carbohydrate diet allows us to provide more energy by fermentation, without using our mitochondria as heavily. But if you have a mitochondrial disease (and we all develop faulty mitochondria with age) then the switch can be too much. Some patients with mitochondrial disease who have adopted a 'ketogenic diet' have collapsed into a coma because their damaged mitochondria can't provide the energy needed for normal life without help from fermentation.

capacity, fertility and longevity. But there is inevitably a limit to what can be achieved by modulating our own physiology. We have a maximum lifespan set by our own evolutionary history, which ultimately depends on the complexity of synaptic connections in our brains, and the size of stem cell populations in other tissues. Henry Ford is said to have visited a scrap heap to determine which parts of derelict Fords were still operational, then insisting that, in new models, these pointlessly long-lasting parts should be replaced with cheaper versions to save money. Likewise, in evolution, there is no point in maintaining a large and dynamic population of stem cells in the stomach lining if they are never used, because our brains wear out first. In the end, we are optimised by evolution to our expected lifespan. I doubt we will ever find a way of living much beyond 120 by merely fine-tuning our physiology.

But evolution is a different matter. Think back to the variable threshold of death. Species with high aerobic requirements, like bats and birds, have a low threshold: even modest free-radical leak will trigger apoptosis during embryonic development, and only the offspring with low leak will develop to full term. This low rate of free-radical leak corresponds to a long lifespan, for the reasons we've just discussed. Conversely, animals with low aerobic requirements – mice, rats, and so on – have a higher death threshold, tolerate higher levels of free-radical leak, and ultimately have shorter lives. There is a straightforward prediction here: selection for greater aerobic capacity, over generations, should prolong lifespan. And so it does. Rats, for example, can be selected for their capacity to run on a treadmill. If the highest-capacity runners in each generation are mated among themselves, and the same is done with the lowest-capacity runners, lifespan increases in the high-capacity group and decreases in the low-capacity group. Over ten generations, the high-capacity runners increase their aerobic capacity by 350% relative to the low-capacity runners, and live nearly a year longer (a large difference, given that rats normally live for about three years). I would argue that similar selection took place during the evolution of bats and birds, and indeed endotherms (warm-blooded animals) more generally, ultimately increasing their lifespan by an order of magnitude.[12]

---

12 I discuss the interplay between aerobic capacity and the evolution of endothermy in

We may not wish to select ourselves on such a basis; that smacks far too much of eugenics. Even if it actually worked, such social engineering would produce more problems than it solves. But in fact we might already have done so. We do have a high aerobic capacity relative to other great apes. We do live a lot longer than them – we live nearly twice as long as chimps and gorillas, which have a similar metabolic rate. Perhaps we owe that to our formative years as a species, chasing gazelle across the African savannah. You might not gain enormous pleasure from endurance running, but it sculpted us as a species. There's no gain without pain. From a simple consideration of the requirements for two genomes, we can predict that our ancestors increased their aerobic capacity, decreased their free-radical leak, gave themselves a problem with fertility, and increased their lifespan. How much truth is there in all that? This is a testable hypothesis that might prove wrong. But it emerges inexorably from the requirement for mosaic mitochondria, a prediction that in turn rests on the origin of the eukaryotic cell, which on one singular occasion, nearly 2 billion years ago, overcame the energetic constraints that keep bacteria bacterial. No wonder the sun setting over the plains of Africa still holds such a strong emotional resonance. It ties us by a wonderful, if contorted, train of causality right back to the very origins of life on our planet.

---

some detail in *Power, Sex, Suicide* and *Life Ascending*. I can only shamelessly recommend them, if you want to know more about that.

EPILOGUE: FROM THE DEEP

M̲ore than 1,200 metres deep in the Pacific Ocean off the coast of Japan lies an underwater volcano named Myojin Knoll. A team of Japanese biologists have been trawling these waters for more than a decade, searching for interesting life forms. By their own account, they didn't find anything terribly surprising until May 2010, when they collected some polychaete worms clinging to a hydrothermal vent. It wasn't the worms that were interesting but the microbes associated with them. Well, one of the microbes – one cell that looked a lot like a eukaryote, until they looked at it more closely (**Figure 37**). Then it became the most teasing enigma.

Eukaryote means 'true nucleus', and this cell has a structure that on first glance looks like a normal nucleus. It also has other convoluted internal membranes, and some endosymbionts that could be hydrogenosomes, derived from mitochondria. Like eukaryotic fungi and algae, it has a cell wall; and not surprisingly, for a specimen from the deep black ocean, it lacks chloroplasts. The cell is modestly large, around 10 micrometres in length and 3 micrometres in diameter, giving it a volume about 100 times larger than a typical bacterium such as *E. coli*. The nucleus is large, taking up nearly half of the volume of the cell. On a quick glance, then, this cell isn't easy to classify into a known group, but it is plainly eukaryotic. It's only a matter of time and gene sequencing, you might think, before it is safely assigned to its proper home in the tree of life.

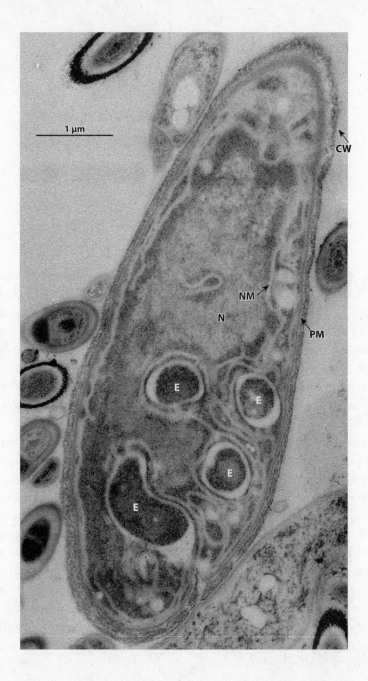

Figure 37 **A unique microorganism from the deep sea**
Is this a prokaryote or a eukaryote? It has a cell wall (CW), plasma membrane
(PM) and a nucleus (N) surrounded by a nuclear membrane (NM). It also has
several endosymbionts (E) which look a bit like hydrogenosomes. It's quite big,
about 10 micrometres in length, and the nucleus is large, taking up nearly 40% of
the cell volume. Plainly a eukaryote, then. But no! The nuclear membrane is a
single layer, not a double membrane. There are no nuclear pore complexes, just
occasional gaps. There are ribosomes in the nucleus (mottled grey regions) and
outside the nucleus. The nuclear membrane is continuous with other membranes
and even the plasma membrane. The DNA is in the form of thin filaments, 2
nanometres in diameter as in bacteria, not eukaryotic chromosomes. Plainly not a
eukaryote, then. I suspect this enigma is actually a prokaryote that acquired
bacterial endosymbionts, and is now recapitulating eukaryotic evolution,
becoming larger, swelling its genome, accumulating the raw material for
complexity. But this is the only sample, and without a genome sequence we may
never know.

Oh, but look again! All eukaryotes have a nucleus, true, but in all known
cases that nucleus is similar in its structure. It has a doubled membrane,
continuous with other cellular membranes, a nucleolus, where ribosomal
RNA is synthesised, elaborate nuclear pore complexes, and an elastic
lamina; and the DNA is carefully packaged in proteins, forming chromosomes – relatively thick chromatin fibres, 30 nanometres in diameter. As we
saw in Chapter 6, protein synthesis takes place on ribosomes that are always
excluded from the nucleus. This is the very basis of the distinction between
the nucleus and cytoplasm. So what about the cell from Myojin Knoll? It has
a single nuclear membrane, with a few gaps. No nuclear pores. The DNA
is composed of fine fibres as in bacteria, about 2 nanometres in diameter, not
thick eukaryotic chromosomes. There are ribosomes in the nucleus. Ribosomes in the nucleus! And ribosomes outside the nucleus too. The nuclear
membrane is continuous with the cell membrane in several places. The
endosymbionts could be hydrogenosomes, but some of them have a bacterial corkscrew morphology on 3D reconstruction. They look more like relatively recent bacterial acquisitions. While it has internal membranes there is
nothing resembling an endoplasmic reticulum, or the Golgi apparatus, or a

cytoskeleton, all classic eukaryotic traits. In other words, this cell is actually nothing like a modern eukaryote. It just bears a superficial resemblance.

So what is it, then? The authors didn't know. They named the beast *Parakaryon myojinensis*, the new term 'parakaryote' signifying its intermediate morphology. Their paper, published in the *Journal of Electron Microscopy*, had one of the most tantalising titles I've ever seen: 'Prokaryote or eukaryote? A unique microorganism from the deep sea.' Having set up the question beautifully, the paper goes nowhere at all in answering it. A genome sequence, or even a ribosomal RNA signature, would give some insight into the true identity of the cell, and turned this largely overlooked scientific footnote into a high-impact *Nature* paper. But they had sectioned their only sample. All they can say for sure is that in 15 years and 10,000 electron microscopy sections, they had never seen anything remotely similar before. They haven't seen anything similar since, either. Neither has anyone else.

So what is it, then? The unusual traits could be an artefact of preparation – a possibility that is not to be discounted, given the troubled history of electron microscopy. On the other hand, if the traits are just an artefact, why was this sample a unique oddity? And why do the structures look so reasonable in themselves? I'd hazard it's not an artefact. That leaves three conceivable alternatives. It could be a highly derived eukaryote, which changed its normal structures as it adapted to an unusual lifestyle, clinging to the back of a deep-sea worm on a hydrothermal vent. But that seems unlikely. Plenty of other cells live in similar circumstances, and they have not followed suit. In general, highly derived eukaryotes lose archetypal eukaryotic traits, but those that remain are still recognisably eukaryotic. That's true of all the archezoa, for example, those purportedly living fossils that were once thought to be primitive intermediates but eventually turned out to be derived from fully fledged eukaryotes. If *Parakaryon myojinensis* really is a highly derived eukaryote, then it's radically different in its basic plan to anything we've seen before. I don't think that's what it is.

Alternatively, it could be a real living fossil, a 'genuine archezoan' that somehow clung to existence, failing to evolve the modern range of eukaryotic accessories in the unchanging deep oceans. This explanation is favoured by the authors of the paper, but I don't believe that either. It is not living in an unchanging environment: it is attached to the back of a polychaete worm, a complex multicellular eukaryote that obviously did not exist in the early evolution of eukaryotes. The low population density – just a single cell discovered after many years of trawling – also makes me doubt that it could have survived unchanged for nearly 2 billion years. Small populations are highly prone to extinction. If the population expands, fine; but if not, it's only a matter of time before random statistical chance pushes it into obliv- ion. Two billion years is a very long time – about 30 times longer than the period coelacanths are thought to have survived as living fossils in the deep oceans. Any genuine survivors from the early days of the eukaryotes would have to be at least as populous as the real archezoa to survive that long.

That leaves the final possibility. As Sherlock Holmes remarked, 'When you have eliminated all which is impossible, then whatever remains, however improbable, must be the truth.' While the other two options are by no means impossible, this third is much the most interesting: it is a prokaryote, which has acquired endosymbionts, and is changing into a cell that resembles a eukaryote, in some kind of evolutionary recapitulation. To my mind, that makes much more sense. It immediately explains why the population density is low; as we've seen, endosymbioses between prokaryotes are rare and are beset by logistical difficulties.[1] It's not at all easy to reconcile selection acting at the level of the host cell and the endosymbiont in a 'virgin' endosymbiosis between prokaryotes. The most likely fate for this cell is extinction. An endosymbiosis between prokaryotes also explains why this cell has various traits that look eukaryotic, but on closer inspection are not. It is relatively large, with a genome that looks

---

1 The endosymbionts in *Parakaryon myojinensis* are found inside what the authors describe as phagosomes (vacuoles in the cell) despite the presence of an intact cell wall. They conclude that the host cell must once have been a phagocyte, but later lost this capacity. That's not necessarily the case. Look again at **Figure 25**. These intracellular bacteria are enclosed by very similar 'vacuoles', but in this case the host cell is recognisably a cyanobacterium, and therefore not phagocytic. Dan Wujek ascribes these vacuoles surrounding the endosymbionts to shrinkage during preparation for electron microscopy, and I'd guess the 'phagosomes' in *Parakaryon myojinensis* are also an artefact of shrinkage, and have nothing to do with phagocytosis. If so, there's no reason to think that the ancestral host cell was a more complex phagocyte.

substantially larger than any other prokaryote, housed in a 'nucleus' continuous with internal membranes, and so on. These are all traits that we predicted would evolve, from first principles, in prokaryotes with endosymbionts.

I would wager a small bet that these endosymbionts have already lost a large part of their genome, as I have argued that only the process of endosymbiotic gene loss can support the expansion of the host-cell genome up to eukaryotic levels. That seems to be happening here: an equivalent extreme genomic asymmetry is supporting an independent origin of morphological complexity. Certainly the host-cell genome is large, occupying more than a third of a cell that is already 100 times larger than *E. coli*. This genome is housed in a structure that looks superficially much like a nucleus. Oddly, only some of the ribosomes are excluded from this structure. Does that mean that the intron hypothesis is wrong? It's hard to say, as the host cell here could be a bacterium, not an archaeon, and so could be less vulnerable to the transfer of bacterial mobile introns. The fact that a nuclear compartment has evolved independently would tend to suggest that similar forces are operating here, and by the same token would tend to operate in large cells with endosymbionts. What about other eukaryotic traits such as sex, and mating types? We simply can't say, without a genome sequence. As I noted, this really is the most teasing enigma. We'll just have to wait and see; that's part and parcel of the never-ending uncertainty of science.

This whole book has been an attempt to predict why life is the way it is. To a first approximation, it looks as if *Parakaryon myojinensis* might be recapitulating a parallel pathway towards complex life, from bacterial ancestors. Whether that same pathway is followed elsewhere in the universe hinges on the starting point – the origin of life itself. I have argued that this starting point might well be recapitulated too.

All life on earth is chemiosmotic, depending on proton gradients across membranes to drive carbon and energy metabolism. We have explored the possible origins and consequences of this peculiar trait. We've seen that living requires a continuous driving force, an unceasing chemical reaction that produces reactive intermediates, including molecules like ATP, as by-products. Such molecules drive the energy-demanding reactions that make up cells. This flux of carbon and energy must have been even greater at the origins of life, before the evolution of biological catalysts, which constrained the flow of metabolism within narrow channels. Very few natural environments meet the requirements for life – a continuous, high flux of carbon and usable energy across mineral catalysts, constrained in a naturally microcompartmentalised system, capable of concentrating products and venting waste. While there may be other environments that meet these criteria, alkaline hydrothermal vents most certainly do, and such vents are likely to be common on wet rocky planets across the universe. The shopping list for life in these vents is just rock (olivine), water and CO_2, three of the most ubiquitous substances in the universe. Suitable conditions for the origin of life might be present, right now, on some 40 billion planets in the Milky Way alone.[2]

Alkaline hydrothermal vents come with both a problem and a solution: they are rich in H_2, but this gas does not react readily with CO_2. We have seen that natural proton gradients across thin semiconducting mineral barriers could theoretically drive the formation of organics, and ultimately the emergence of cells, within the pores of the vents. If so, life depended from the very beginning on proton gradients (and iron–sulphur minerals) to break down the kinetic barriers to the reaction of H_2 and CO_2. To grow on natural proton gradients, these early cells required leaky membranes, capable of retaining the molecules needed for life without cutting themselves off from the energising flux of protons. That, in turn, precluded their escape from the vents, except through the strait gates of a strict succession of events (requiring an antiporter), which enabled the coevolution of active ion pumps and modern phospholipid membranes. Only then could cells leave the vents, and colonise the oceans and rocks of the early earth. We saw that this strict succession of events could explain the paradoxical properties of LUCA, the last universal common ancestor of life, as well as the deep divergence of bacteria and archaea. Not least, these strict requirements can explain why all life on earth is chemiosmotic – why this strange trait is as universal as the genetic code itself.

---

2 Data from the space telescope Kepler suggests that 1 in 5 sun-like stars in the galaxy have an 'earth-sized' planet in the habitable zone, giving that projected total of 40 billion suitable planets in the Milky Way.

This scenario – an environment that is common in cosmic terms, but with a strict set of constraints governing outcomes – makes it likely that life elsewhere in the universe will also be chemiosmotic, and so will face parallel opportunities and constraints. Chemiosmotic coupling gives life unlimited metabolic versatility, allowing cells to 'eat' and 'breathe' practically anything. Just as genes can be passed around by lateral gene transfer, because the genetic code is universal, so too the toolkit for metabolic adaptation to very diverse environments can be passed around, as all cells use a common operating system. I would be amazed if we did not find bacteria right across the universe, including our own solar system, all working in much the same way, powered by redox chemistry and proton gradients across membranes. It's predictable from first principles.

But if that's true, then complex life elsewhere in the universe will face exactly the same constraints as eukaryotes on earth – aliens should have mitochondria too. We've seen that all eukaryotes share a common ancestor which arose just once, through a rare endosymbiosis between prokaryotes. We know of two such endosymbioses between bacteria (**Figure 25**) – three, if we include *Parakaryon myojinensis* – so we know that it is possible for bacteria to get inside bacteria without phagocytosis. Presumably there must have been thousands, perhaps millions, of cases over 4 billion years of evolution. It's a bottleneck, but not a stringent one. In each case, we would expect to see gene loss from the endosymbionts, and a tendency to greater size and genomic complexity in the host cell – exactly what we do see in *Parakaryon myojinensis*. But we'd also expect intimate conflict between the host and the endosymbiont – this is the second part of the bottleneck, a double whammy that makes the evolution of complex life genuinely difficult. We saw that the first eukaryotes most likely evolved quickly in small populations; the very fact that the common ancestor of eukaryotes shares so many traits, none of which are found in bacteria, implies a small, unstable, sexual population. If *Parakaryon myojinensis* is recapitulating eukaryotic evolution, as I suspect, its extremely low population density (just one specimen in 15 years of hunting) is predictable. Its most likely fate is extinction. Perhaps it will die because it has not successfully excluded all its ribosomes from its nuclear compartment, or because it has not yet 'invented' sex. Or perhaps, chance in a million, it will succeed, and seed a second coming of eukaryotes on earth.

I think we can reasonably conclude that complex life will be rare in the universe – there is no innate tendency in natural selection to give rise to humans or any other form of complex life. It is far more likely to get stuck at the bacterial level of complexity. I can't put a statistical probability on that. The existence of *Parakaryon myojinensis* might be encouraging for some – multiple origins of complexity on earth means that complex life might be more common elsewhere in the universe. Maybe. What I would argue with more certainty is that, for energetic reasons, the evolution of complex life requires an endosymbiosis between two prokaryotes, and that is a rare random event, disturbingly close to a freak accident, made all the more difficult by the ensuing intimate conflict between cells. After that, we are back to standard natural selection. We've seen that many properties shared by eukaryotes, from the nucleus to sex, are predictable from first principles. We can go much further. The evolution of two sexes, the germ-line–soma distinction, programmed cell death, mosaic mitochondria, and the trade-offs between aerobic fitness and fertility, adaptability and disease, ageing and death, all these traits emerge, predictably, from the starting point that is a cell within a cell. Would it all happen over again? I think that much of it would. Incorporating energy into evolution is long overdue, and begins to lay a more predictive basis to natural selection.

Energy is far less forgiving than genes. Look around you. This wonderful world reflects the power of mutations and recombination, genetic change – the basis for natural selection. You share some of your genes with the tree through the window, but you and that tree parted company very early in eukaryotic evolution, 1.5 billion years ago, each following a different course permitted by different genes, the product of mutations, recombination, and natural selection. You run around, and I hope still climb trees occasionally; they bend gently in the breeze and convert the air into more trees, the magic trick to end them all. All of those differences are written in the genes, genes that derive from your common ancestor but have now mostly diverged beyond recognition. All those changes were permitted, selected, in the long course of evolution. Genes are almost infinitely permissive: anything that can happen will happen.

But that tree has mitochondria too, which work in much the same way as its chloroplasts, endlessly transferring electrons down its trillions upon trillions of respiratory chains, pumping protons across membranes as they always did. As you always did. These same shuttling electrons and protons have sustained you from the womb: you pump 10^{21} protons per second, every second, without pause. Your mitochondria were passed on from your mother, in her egg cell, her most precious gift, the gift of living that goes back unbroken, unceasing, generation on generation, to the first stirrings of life in hydrothermal vents, 4 billion years ago. Tamper with this reaction at your peril. Cyanide will stem the flow of electrons and protons, and bring your life to an abrupt end. Ageing will do the same, but slowly, gently. Death is the ceasing of electron and proton flux, the settling of membrane potential, the end of that unbroken flame. If life is nothing but an electron looking for a place to rest, death is nothing but that electron come to rest.

This energy flux is astonishing and unforgiving. Any change over seconds or minutes could bring the whole experiment to an end. Spores can pull it off, descending into metabolic dormancy from which they must feel lucky to emerge. But for the rest of us … we are sustained by the same processes that powered the first living cells. These processes have never changed in a fundamental way; how could they? Life is for the living. Living needs an unceasing flux of energy. It's hardly surprising that energy flux puts major constraints on the path of evolution, defining what is possible. It's not surprising that bacteria keep doing what bacteria do, unable to tinker in any serious way with the flame that keeps them growing, dividing, conquering. It's not surprising that the one accident that did work out, that singular endosymbiosis between prokaryotes, did not tinker with the flame, but ignited it in many copies in each and every eukaryotic cell, finally giving rise to all complex life. It's not surprising that keeping this flame alive is vital to our physiology and evolution, explaining many quirks of our past and our lives today. How lucky that our minds, the most improbable biological machines in the universe, are now a conduit for this restless flow of energy, that we can think about why life is the way it is. May the proton-motive force be with you!

GLOSSARY

aerobic respiration – our own form of respiration, in which energy from the reaction between food and oxygen is harnessed to power work; bacteria can also 'burn' minerals or gases with oxygen. See also *anaerobic respiration* and *respiration*.

alkaline hydrothermal vent – a type of vent, usually on the sea floor, that emits warm alkaline fluids rich in hydrogen gas; probably played a major role in the origin of life.

allele – one particular form of a gene in a population.

amino acid – one of 20 distinct molecular building blocks that are linked together in a chain to form a protein (often containing hundreds of amino acids).

anaerobic respiration – any one of many alternative forms of respiration, common in bacteria, in which molecules other than oxygen (such as nitrate or sulphate) are used to 'burn' (oxidise) food, minerals or gases. **Anaerobes** are organisms that live without oxygen. See also *aerobic respiration* and *respiration*.

ångström (Å) – a unit of distance, roughly the scale of an atom, technically one ten-billionth of a metre (10^{-10} m); a **nanometre** is 10 times that length, one-billionth of a metre (10^{-9} m).

antiporter – a protein 'turnstile' that typically exchanges one charged atom (ion) for another across a membrane, for example a proton (H^+) for a sodium ion (Na^+).

apoptosis – 'programmed' cell death, an energy-consuming process encoded by genes, in which a cell dismantles itself.

archaea – one of the three great domains of life, the other two being the bacteria and eukaryotes (such as ourselves); archaea are prokaryotes, lacking a nucleus to store their DNA, and most other elaborate structures found in complex eukaryotes.

archezoa – don't mix these up with archaea! Archezoa are simple, single-celled eukaryotes, once mistaken for evolutionary 'missing links' between bacteria and more complex eukaryotic cells.

ATP – adenosine triphosphate, the biological energy 'currency' used by all known cells. **ADP** (adenosine diphosphate) is the breakdown product formed when ATP is 'spent'; the energy of respiration is used to join a phosphate (PO_4^{3-}) back on to ADP to reform ATP. **Acetyl phosphate** is a simple (two-carbon) biological energy 'currency' that works a bit like ATP, which could have been formed by geological processes on the early earth.

ATP synthase – a remarkable rotating motor protein, a nanoturbine that sits in the membrane and uses the flow of protons to power the synthesis of ATP.

bacteria – one of the three great domains of life, the other two being the archaea and eukaryotes (like us); along with archaea, bacteria are prokaryotes, which lack a nucleus to store their DNA, as well as most other elaborate structures found in complex eukaryotes.

chemiosmotic coupling – the way in which energy from respiration is used to pump protons across a membrane; the flux of protons back through protein turbines in the membrane (ATP synthase) then drives the formation of ATP. So respiration is 'coupled' to ATP synthesis by a proton gradient.

chloroplast – a specialised compartment in plant cells and algae where photosynthesis takes place; originally derived from photosynthetic bacteria called cyanobacteria.

chromosome – a tubular structure composed of DNA tightly wrapped in proteins, visible during cell division; humans have 23 pairs of distinct chromosomes containing two copies of all our genes. A **fluid chromosome** undergoes recombination, giving different combinations of genes (alleles).

cytoplasm – the gel-like substance of cells, excluding the nucleus; the **cytosol** is the watery solution surrounding internal compartments such as mitochondria. The **cytoskeleton** is the dynamic protein scaffold inside cells, which can form and reform as cells change shape.

disequilibrium – a potentially reactive state in which molecules that 'want' to react with each other have yet to do so. Organic matter and oxygen are in disequilibrium – given the opportunity (striking a match) organic matter will burn.

dissipative structure – a stable physical structure that takes a characteristic form, as in a whirlpool, hurricane, or the jet stream, sustained by a continuous flux of energy.

DNA – deoxyribonucleic acid, the hereditary material, which takes the form of a double helix; **parasitic DNA** is DNA that can copy itself selfishly, at the expense of the individual organism.

electron – a subatomic particle that carries a negative electrical charge. An **electron acceptor** is an atom or molecule that gains one or more electrons; an **electron donor** loses electrons.

endergonic – a reaction that requires an input of free energy ('work', not heat) to proceed. An **endothermic** reaction requires an input of heat to proceed.

endosymbiosis – a mutual relationship (usually a trade of metabolic substances) between two cells, in which one partner physically lives inside the other.

entropy – a state of molecular disorder tending towards chaos.

enzyme – a protein that catalyses a particular chemical reaction, often increasing its rate by millions of times the uncatalysed rate.

eukaryote – any organism composed of one or more cells containing a nucleus and other specialised structures like mitochondria; all complex life forms, including plants, animals, fungi, algae, and protists such as the amoeba, are made of eukaryotic cells. Eukaryotes form one of the three great domains of life, the other two being the simpler prokaryotic domains of bacteria and archaea.

exergonic – a reaction that releases free energy, which can power work. An **exothermic** reaction releases heat.

fatty acid – a long-chain hydrocarbon, typically with 15–20 linked carbon atoms, used in the fatty (lipid) membranes of bacteria and eukaryotes; always has an acid group at one end.

fermentation – this is *not* anaerobic respiration! Fermentation is a purely chemical process of generating ATP, which does not involve proton gradients across membranes or the ATP synthase. Different organisms have slightly different pathways; we produce lactic acid as a waste product, yeasts form alcohol.

FeS cluster – iron–sulphur cluster, a small mineral-like crystal composed of a lattice of iron and sulphur atoms (usually the compound Fe_2S_2 or Fe_4S_4) found at the heart of many important proteins, including some of those used in respiration.

fixation – when one particular form of a gene (allele) is found in all individuals in a population.

free energy – energy that is free to power work (not heat).

free radical – an atom or molecule with an unpaired electron (which tends to make it unstable and reactive); oxygen free radicals escaping from respiration may play a role in ageing and disease.

gene – a stretch of DNA encoding a protein (or other product such as regulatory RNA). The **genome** is the total compendium of genes in an organism.

germline – the specialised sex cells in animals (such as sperm and egg cells), which alone pass on the genes that give rise to new individuals in each generation.

intron – a 'spacer' sequence within a gene, which does not code for a protein and is usually removed from the code-script before the protein is made. **Mobile introns** are genetic parasites that can copy themselves repeatedly within a genome; eukaryotic introns apparently derive from a proliferation of mobile bacterial introns early in eukaryotic evolution, followed by mutational decay.

lateral gene transfer – the transfer of (usually) a small number of genes from one cell to another or the uptake of naked DNA from the environment. Lateral gene transfer is a trade in genes within the same generation; in **vertical inheritance**, the entire genome is copied and passed on to the daughter cells at cell division.

LUCA – the last universal common ancestor of all cells living today, whose hypothetical properties can be reconstructed by comparing the properties of modern cells.

meiosis – the process of reductive cell division in sex, to form gametes which have a single complete set of chromosomes (making it haploid) rather than the two sets found in the parent cells (diploid). **Mitosis** is the normal form of cell division in eukaryotes, in which chromosomes are doubled, then separated into two daughter cells on a microtubular spindle.

membrane – a very thin fatty layer surrounding cells (also found inside cells); composed of a 'lipid bilayer' with a hydrophobic (water-hating) interior and hydrophilic (water-loving) head-groups on either side. **Membrane potential** is the electrical charge (potential difference) between opposite sides of the membrane.

metabolism – the set of life-sustaining chemical reactions within living cells.

mitochondria – the discrete 'powerhouses' in eukaryotic cells, which derive from α-proteobacteria, and which retain a tiny but hugely important genome of their own. **Mitochondrial genes** are those physically located within the mitochondria. **Mitochondrial biogenesis** is the replication, or growth, of new mitochondria, which also requires genes in the nucleus.

monophyletic radiation – the divergence of multiple species from a single common ancestor (or a single phylum) like the spokes of a wheel radiating from a central hub.

mutation – usually refers to a change in the specific sequence of a gene, but can also include other genetic changes, such as random deletions or duplications of DNA.

nucleus – the 'control centre' of complex (eukaryotic) cells, which contains most of the cell's genes (some are found in mitochondria).

nucleotide – one of the building blocks linked together in a chain to form RNA and DNA; there are scores of related nucleotides that act as cofactors in enzymes, catalysing specific reactions.

ortholog – the same gene with the same function found in different species, all of which inherited it from a common ancestor.

oxidation – the removal of one or more electrons from a substance, rendering it **oxidised.**

paralog – a member of a family of genes, formed by gene duplications within the same genome; equivalent gene families can also be found in different species, inherited from a common ancestor.

pH – a measure of acidity, specifically the concentration of protons: acids have a high concentration of protons (giving them a low pH, below 7); alkalis have a low concentration of protons, giving them a high pH (7–14); pure water has a neutral pH (7).

phagocytosis – the physical engulfing of one cell by another, swallowing it up into a 'food' vacuole to be digested internally. **Osmotrophy** is the external digestion of food, followed by absorption of small compounds, as practised by fungi.

photosynthesis – the conversion of carbon dioxide into organic matter, using solar energy to extract electrons from water (or other substances) and ultimately attach them on to carbon dioxide.

plasmid – a small ring of parasitic DNA that transmits selfishly from one cell to another; plasmids can also provide useful genes for their host cells (such as genes conferring antibiotic resistance).

polyphyletic radiation – the divergence of multiple species from a number of evolutionarily distinct ancestors (different phyla) like the spokes of multiple wheels radiating from multiple hubs.

prokaryote – a general term connoting simple cells that lack a nucleus (literally 'before the nucleus') and including both bacteria and archaea, two of the three domains of life.

protein – a chain of amino acids linked together in a precise order specified by the sequence of DNA letters in a gene; a **polypeptide** is a shorter chain of amino acids, whose order need not be specified.

protist – any single-celled eukaryote, some of which can be very complex, with as many as 40,000 genes and an average size at least 15,000 times larger than bacteria; **protozoa** is a vivid but defunct term (meaning 'first animals') which referred to protists such as the amoeba that behave like animals.

proton – a subatomic particle with positive charge; a hydrogen atom is composed of a single proton and a single electron; loss of the electron leaves the hydrogen nucleus, the positively charged proton denoted H^+.

proton gradient – a difference in the concentration of protons on opposite sides of a membrane; the **proton-motive force** is the electrochemical force resulting from the combined difference in electrical charge and concentration of H^+ across a membrane.

recombination – the exchange of one piece of DNA for an equivalent piece from another source, giving rise to different combinations of genes (specifically alleles) on 'fluid' chromosomes.

redox – the combined process of reduction and oxidation, which amounts to a transfer of electrons from a donor to an acceptor. A **redox couple** is a specific electron donor with a specific acceptor; a **redox centre** receives an electron before passing it on, making it both an acceptor and a donor.

reduction – the addition of one or more electrons to a substance, rendering it **reduced**.

replication – the duplication of a cell or molecule (typically DNA) to give two daughter copies.

respiration – the process by which nutrients are 'burned' (oxidised) to generate energy in the form of ATP. Electrons are stripped from food or other electron donors (such as hydrogen) and passed on to oxygen or other oxidants (such as nitrate) via a series of steps called the **respiratory chain**. The energy released is used to pump protons across a membrane, generating a proton-motive force that in turn drives ATP synthesis. See also *anaerobic respiration* and *aerobic respiration*.

ribosome – protein-building 'factories' found in all cells, which convert the RNA code-script (copied from DNA) into a protein with the correct sequence of amino acid building blocks.

RNA – ribonucleic acid; a close cousin of DNA, but with two tiny chemical alterations that transform its structure and properties. RNA is found in three main forms: messenger RNA (a code-script copied from DNA); transfer RNA (which delivers amino acids according to the genetic code); and ribosomal RNA (which acts as 'machine parts' in ribosomes).

RNA world – a hypothetical early stage of evolution in which RNA acts simultaneously as a template for its own replication (in place of DNA) and a catalyst that speeds up reactions (in place of proteins).

selective sweep – strong selection for a particular genetic variant (allele), eventually displacing all other variants from a population.

selfish conflict – a metaphorical clash between the interests of two distinct entities, for example between endosymbionts or plasmids and a host cell.

serpentinisation – a chemical reaction between certain rocks (minerals rich in magnesium and iron, such as olivine) and water, giving rise to strongly alkaline fluids saturated in hydrogen gas.

sex – a reproductive cycle, involving the division of cells by meiosis to form gametes, each with half the normal quota of chromosomes, followed by the fusion of gametes to produce a fertilised egg.

sex determination – the processes controlling male or female development.

snowball earth – a global freeze, with glaciers encroaching to sea level at the equator; thought to have occurred on several occasions in earth's history.

substrate – substances required for cell growth, converted by enzymes into biological molecules.

thermodynamics – a branch of physics dealing with heat, energy and work; thermodynamics governs the reactions that could occur under a particular set of conditions; **kinetics** defines the rate at which such reactions actually do take place.

thermophoresis – the concentration of organics by thermal gradients or convection currents.

transcription – the formation of a short RNA code-script (called messenger RNA) from DNA, as the first step in making a new protein.

translation – the physical assembly of a new protein (on a ribosome) in which the precise sequence of amino acids is specified by an RNA code-script (messenger RNA).

uniparental inheritance – the systematic inheritance of mitochondria from only one of two parents, typically from the egg but not the sperm; **biparental inheritance** is the inheritance of mitochondria from both parents.

variance – a measure of spread in a set of numbers; if variance is zero, all the values are identical; if variance is small, the values all fall close to the mean; high variance indicates a wide range of values.

ACKNOWLEDGEMENTS

This book marks the end of a long personal journey, and the beginning of a new journey. The first journey began as I was writing an earlier book, *Power, Sex, Suicide: Mitochondria and the Meaning of Life*, published by OUP in 2005. That was when I first began to grapple with the questions I deal with here – the origins of complex life. I was strongly influenced by the extraordinary work of Bill Martin on the origin of the eukaryotic cell, and his equally radical work with pioneering geochemist Mike Russell on the origin of life and the very early divergence of archaea and bacteria. Everything in that (and this) book is grounded in the context laid out by these two colossi of evolutionary biology. But some of the ideas developed here are original. Writing books gives one scope to think, and for me this is an unmatchable pleasure of writing for a general audience: I have to think clearly, to try to express myself in a way that, first of all, I can understand myself. That brings me face to face with the things I don't understand, some of which, thrillingly, turn out to reflect universal ignorance. So *Power, Sex, Suicide* was more or less obliged to put forward a few original ideas, which I have been living with ever since.

I presented these ideas at conferences and universities around the world, and grew used to dealing with astute criticisms. My ideas became refined, as did my overall conception of the importance of energy in evolution; and a few cherished thoughts were thrown away as wrong. But however good an idea may be, it only becomes real science when it is framed and tested as a rigorous hypothesis. That seemed a pipe dream until 2008, when University College London announced a new prize for 'ambitious thinkers' to explore paradigm-shifting ideas. The Provost's Venture Research Prize was the brainchild of Professor Don Braben, a dynamo of a man, who has long fought for 'scientific freedom'. Science is fundamentally unpredictable, Braben argues, and can't be constrained to order, however much society might wish to prioritise the spending of tax-payers' money. Genuinely transformative ideas almost always come from the least expected quarters; that alone can be relied upon. Such ideas are transformative not only in their science, but also to the wider economy, which is fuelled by scientific advances. It's therefore in the best interests of society to fund scientists on the strength of their ideas alone, however intangible these might seem, rather than trying to target the perceived benefits to humanity. That rarely works because radically new insights usually come from outside a field altogether; nature has no respect for human boundaries.[1]

Luckily, I was eligible to apply to the UCL scheme. I had a book full of ideas that desperately needed testing, and thankfully, Don Braben was eventually persuaded. While the impetus behind the prize came from Don, to whom I am immensely indebted, I owe as much to the generosity and scientific vision of Professor David Price, Vice Provost for Research at UCL, and the then Provost, Professor Malcolm Grant, for supporting both the scheme and me personally. I'm also extremely grateful to Professor Steve Jones for his support, and for welcoming me to the department he headed at the time: Genetics, Evolution and Environment, natural home for the research I was to undertake.

That was six years ago. Since then, I have been attacking as many of the problems, from as many angles, as I can. The Venture Research funding itself lasted three years, enough to set my direction and give me a fighting

---

1 If you want to know more, Braben has framed his arguments in several compelling books, the latest of which is titled *Promoting the Planck Club: How Defiant Youth, Irreverent Researchers and Liberated Universities Can Foster Prosperity Indefinitely* (Wiley, 2014).

chance of winning funding from other sources to continue. In this I am very grateful to the Leverhulme Trust, who have supported my work on the origin of life over the last three years. Not many organisations are willing to back a genuinely new experimental approach, with all the teething problems that entails. Thankfully, our little bench-top origin-of-life reactor is now beginning to produce exciting results, none of which would have been possible without their support. This book is a distillation of the first meaning from these studies, the beginning of a new journey.

Of course, none of this work has been done alone. I have bounced many ideas back and forth with Bill Martin, Professor of Molecular Evolution at the University of Düsseldorf, who is always generous with his time, energy and ideas, while never hesitating to demolish poor reasoning, or ignorance. It has been a genuine privilege to write several papers with Bill, which I'd like to think are noteworthy contributions to the field. Certainly, few experiences in life can match the intensity and pleasure of writing a paper with Bill. I've learnt another important lesson from Bill – never clutter the problem with possibilities that can be imagined, but are unknown in the real world; focus always on what life as we know it actually does, and then ask why.

I'm equally grateful to Andrew Pomiankowski, Professor of Genetics at UCL, more commonly known as POM. POM is an evolutionary geneticist steeped in the intellectual traditions of the field, having worked with legendary figures such as John Maynard Smith and Bill Hamilton. POM combines their rigour with an eye for the unsolved problems of biology. If I have succeeded in persuading him that the origin of complex cells is just such a problem, he has introduced me to the abstract but powerful world of population genetics. Approaching the origins of complex life from such contrasting points of view has been a steep learning curve, and great fun.

Another good friend at UCL, with boundless ideas, enthusiasm and expertise to drive these projects forward, is Professor Finn Werner. Finn brings another sharply contrasting background to the same questions, that of structural biology, and specifically the molecular structure of the RNA polymerase enzyme, one of the most ancient and magnificent molecular machines, which in itself gives insights into the early evolution of life.

Every conversation and lunch with Finn is invigorating, and I return ready for challenges new.

I've also been privileged to work with a number of gifted PhD students and post-docs, who have driven much of this work forward. These fall into two groups, those working on the real down-dirty chemistry of the reactor, and those bringing their mathematical skills to bear on the evolution of eukaryotic traits. In particular, I thank Dr Barry Herschy, Alexandra Whicher and Eloi Camprubi for their skills in making difficult chemistry actually happen in the lab, and for their shared vision; and Dr Lewis Dartnell, who at the outset helped build the prototype reactor and set these experiments in motion. In this endeavour, I'm also grateful to Julian Evans and John Ward, Professors of Materials Chemistry and Microbiology, respectively, who have offered freely their time, skills and laboratory resources to the reactor project and the co-supervision of students. They have been my comrades in arms through this adventure.

The second group of students and post-docs, working on mathematical modelling, have selected themselves from an unparalleled doctoral training programme at UCL, funded until recently by the Engineering and Physical Sciences Research Council. This programme goes by the smart acronym CoMPLEX, standing improbably for the Centre for Mathematics and Physics in the Life Sciences and Experimental Biology. CoMPLEX students working with POM and me include Dr Zena Hadjivasiliou, Victor Sojo, Arunas Radzvilavicius, Jez Owen, and recently, Drs Bram Kuijper and Laurel Fogarty. All started out with rather vague ideas, and turned them into rigorous mathematical models that give striking insights into how biology really works. It has been an exciting ride, and I've given up trying to predict the outcome. This work began with the inspirational Professor Rob Seymour, who knew more biology than most biologists, while being a formidable mathematician. Tragically, Rob died of cancer in 2012 at the age of sixty-seven. He was loved by a generation of students.

While this book is grounded in the work I've published over the last six years with this wide array of researchers (some twenty-five papers in all; see Further Reading), it reflects a much longer period of thinking and discussion at conferences and seminars, by email and in the pub, all of which has sculpted my views. In particular, I must thank Professor Mike Russell, whose revolutionary ideas on the origin of life inspired a rising generation, and whose tenacity in adversity is a model to us all. Likewise I thank Professor John Allen, whose hypotheses on evolutionary biochemistry have lit up the way. John has also been an outspoken defender of academic freedom, which has recently cost him dearly. I thank Professor Frank Harold, whose synthesis of bioenergetics, cell structure and evolution is laid out in several wonderful books, and whose open-minded scepticism has constantly challenged me to go a little further; Professor Doug Wallace, whose conception of mitochondrial energetics as the central driver of ageing and disease is visionary and inspirational; and Professor Gustavo Barja, who sees so clearly through the dense thickets of misunderstanding about free radicals and ageing, that I always turn first to his view. Not least, I must thank Dr Graham Goddard, whose encouragement and plain speaking many years ago changed the course of my life.

These friends and colleagues are of course but the tip of an iceberg. I can't thank all those who have shaped my thinking in detail, but I am indebted to them all. In random order: Christophe Dessimoz, Peter Rich, Amandine Marechal, Sir Salvador Moncada, Mary Collins, Buzz Baum, Ursula Mittwoch, Michael Duchen, Gyuri Szabadkai, Graham Shields, Dominic Papineau, Jo Santini, Jürg Bähler, Dan Jaffares, Peter Coveney, Matt Powner, Ian Scott, Anjali Goswami, Astrid Wingler, Mark Thomas, Razan Jawdat and Sioban Sen Gupta, all at UCL; Sir John Walker, Mike Murphy and Guy Brown (Cambridge); Erich Gnaiger (Innsbruck); Filipa Sousa, Tal Dagan and Fritz Boege (Dusseldorf); Paul Falkowski (Rutgers); Eugene Koonin (NIH); Dianne Newman and John Doyle (Caltech); James McInerney (Maynooth); Ford Doolittle and John Archibald (Dalhousie); Wolfgang Nitschke (Marseilles); Martin Embley (Newcastle); Mark van der Giezen and Tom Richards (Exeter); Neil Blackstone (Northern Illinois); Ron Burton (Scripps); Rolf Thauer (Marburg); Dieter Braun (Munich); Tonio Enríquez (Madrid); Terry Kee (Leeds); Masashi Tanaka (Tokyo); Masashi Yamaguchi, Chiba; Geoff Hill (Auburn); Ken Nealson and Jan Amend (Southern California); Tom McCollom (Colorado); Chris Leaver and Lee Sweetlove (Oxford); Markus Schwarzländer (Bonn); John Ellis

(Warwick); Dan Mishmar (Ben Gurion); Matthew Cobb and Brian Cox (Manchester); Roberto and Roberta Motterlini (Paris); and Steve Iscoe (Queens, Kingston). Thank you, all.

I'm also very grateful to a handful of friends and family who have read and commented on parts (or all) of this book. In particular, my father, Thomas Lane, who has sacrificed time from writing his own books on history to read much of this book, finessing my use of language throughout; Jon Turney, who has likewise been generous with his time and comments, especially on pitch, while engaged on his own writing projects; Markus Schwarzländer, whose enthusiasm buoyed me through difficult periods; and Mike Carter, who alone among my friends has read and commented, with trenchant wit, on every chapter of every book I have written, occasionally even persuading me to change tack. And although none of them have read this book (yet), I have to thank Ian Ackland-Snow, Adam Rutherford and Kevin Fong for lunches and good conversations in the pub. They know well how important that is for sanity.

I need hardly say that this book has benefited greatly from the expertise of my agent and publishers. I'm very grateful to Caroline Dawnay at United Agents, for believing in this project from the outset; to Andrew Franklin at Profile, whose editorial comments cut straight to the quick, making the book much harder hitting; to Brendan Curry at Norton, who pointed astutely to passages lacking clarity; and Eddie Mizzi, whose sensitive copy editing again reflects his fine judgement and eclectic learning. His interventions have spared more blushes than I care to admit. Many thanks, too, to Penny Daniel, Sarah Hull, Valentina Zanca and the team at Profile, for marshalling this book to press and beyond.

And finally I turn to my family. My wife, Dr Ana Hidalgo, has lived and breathed this book with me, reading every chapter at least twice, and always illuminating the way ahead. I trust her judgement and knowledge more than my own, and what good there is in my writing has evolved under her natural selection. I can think of no better way to spend my life than trying to understand life, but I already know that my own meaning and joy stems from Ana, from our wonderful sons Eneko and Hugo, and from our wider family in Spain, England and Italy. This book was written in the happiest of times.

BIBLIOGRAPHY

This selection is far from a complete bibliography, more of an entrée into the literature; these are the books and papers that have particularly influenced my own thinking over the last decade. I don't always agree with them, but they are always stimulating and worth reading. I have included several of my own papers in each chapter, which give a detailed, peer-reviewed basis for the arguments laid out more broadly in the book. These papers contain comprehensive reference lists; if you are keen to nail down my own more detailed sources, they are the places to look. For the more casual reader, there should be plenty to go at in the books and articles listed here. I've grouped the references by theme in each chapter, alphabetically within each section. A few significant papers are cited more than once, being relevant to more than one section.

Introduction

Leeuwenhoek and the early development of microbiology

Dobell C. *Antony van Leeuwenhoek and his Little Animals*. Russell and Russell, New York (1958).

Kluyver AJ. Three decades of progress in microbiology. *Antonie van Leeuwenhoek* **13**: 1–20 (1947).

Lane N. Concerning little animals: Reflections on Leeuwenhoek's 1677 paper. *Philosophical Transactions Royal Society B*. In press (2015).

Leewenhoeck A. Observation, communicated to the publisher by Mr. Antony van Leewenhoeck, in a Dutch letter of the 9 Octob. 1676 here English'd: concerning little animals by him observed in rain-well-sea and snow water; as also in water wherein pepper had lain infused. *Philosophical Transactions Royal Society B* **12**: 821–31 (1677).

Stanier RY, van Niel CB. The concept of a bacterium. *Archiv fur Microbiologie* **42**: 17–35 (1961).

Lynn Margulis and the serial endosymbiosis theory

Archibald J. *One Plus One Equals One*. Oxford University Press, Oxford (2014).

Margulis L, Chapman M, Guerrero R, Hall J. The last eukaryotic common ancestor (LECA): Acquisition of cytoskeletal motility from aerotolerant spirochetes in the Proterozoic Eon. *Proceedings National Academy Sciences USA* **103**, 13080–85 (2006).

Sagan L. On the origin of mitosing cells. *Journal of Theoretical Biology* **14**: 225–74 (1967).

Sapp J. *Evolution by Association: A History of Symbiosis*. Oxford University Press, New York (1994).

Carl Woese and the three domains of life

Crick FHC. The biological replication of macromolecules. *Symposia of the Society of Experimental Biology*. 12, 138–63 (1958).

Morell V. Microbiology's scarred revolutionary. *Science* **276**: 699–702 (1997).

Woese C, Kandler O, Wheelis ML. Towards a natural system of organisms: Proposal for the domains Archaea, Bacteria, and Eucarya. *Proceedings National Academy Sciences USA* **87**: 4576–79 (1990).

Woese CR, Fox GE. Phylogenetic structure of the prokaryotic domain: The primary kingdoms. *Proceedings National Academy Sciences USA* **74**: 5088–90 (1977).

Woese CR. A new biology for a new century. *Microbiology and Molecular Biology Reviews* **68**: 173–86 (2004).

Bill Martin and the chimeric origin of eukaryotes

Martin W, Müller M. The hydrogen hypothesis for the first eukaryote. *Nature* **392**: 37–41 (1998).

Martin W. Mosaic bacterial chromosomes: a challenge en route to a tree of genomes. *BioEssays* **21**: 99–104 (1999).

Pisani D, Cotton JA, McInerney JO. Supertrees disentangle the chimeric origin of eukaryotic genomes. *Molecular Biology and Evolution* **24**: 1752–60 (2007).

Rivera MC, Lake JA. The ring of life provides evidence for a genome fusion origin of eukaryotes. *Nature* **431**: 152–55 (2004).

Williams TA, Foster PG, Cox CJ, Embley TM. An archaeal origin of eukaryotes supports only two primary domains of life. *Nature* **504**: 231–36 (2013).

Peter Mitchell and chemiosmotic coupling

Lane N. Why are cells powered by proton gradients? *Nature Education* **3**: 18 (2010).

Mitchell P. Coupling of phosphorylation to electron and hydrogen transfer by a chemi-osmotic type of mechanism. *Nature* **191**: 144–48 (1961).

Orgell LE. Are you serious, Dr Mitchell? *Nature* **402**: 17 (1999).

Chapter 1: What is Life?

Probability and properties of life

Conway-Morris SJ. *Life's Solution: Inevitable Humans in a Lonely Universe*. Cambridge University Press, Cambridge (2003).

de Duve C. *Life Evolving: Molecules, Mind, and Meaning*. Oxford University Press, Oxford (2002).

de Duve. *Singularities: Landmarks on the Pathways of Life*. Cambridge University Press, Cambridge (2005).

Gould SJ. *Wonderful Life. The Burgess Shale and the Nature of History*. WW Norton, New York (1989).

Maynard Smith J, Szathmary E. *The Major Transitions in Evolution*. Oxford University Press, Oxford. (1995).

Monod J. *Chance and Necessity*. Alfred A. Knopf, New York (1971).

Beginnings of molecular biology

Cobb M. 1953: When genes became information. *Cell* **153**: 503–06 (2013).

Cobb M. *Life's Greatest Secret: The Story of the Race to Crack the Genetic Code*. Profile, London (2015).

Schrödinger E. *What is Life?* Cambridge University Press, Cambridge (1944).

Watson JD, Crick FHC. Genetical implications of the structure of deoxyribonucleic acid. *Nature* **171**: 964–67 (1953).

Genome size and structure

Doolittle WF. Is junk DNA bunk? A critique of ENCODE. *Proceedings National Academy Sciences USA* **110**: 5294–5300 (2013).

Grauer D, Zheng Y, Price N, Azevedo RBR, Zufall RA, Elhaik E. On the immortality of television sets: "functions" in the human genome according to the evolution-free gospel of ENCODE. *Genome Biology and Evolution* **5**: 578–90 (2013).

Gregory TR. Synergy between sequence and size in large-scale genomics. *Nature Reviews Genetics* **6**: 699–708 (2005).

First two billion years of life on Earth

Arndt N, Nisbet E. Processes on the young earth and the habitats of early life. *Annual Reviews Earth and Planetary Sciences* **40**: 521–49 (2012).

Hazen R. *The Story of Earth: The First 4.5 Billion Years, from Stardust to Living Planet.* Viking, New York (2014).

Knoll A. *Life on a Young Planet: The First Three Billion Years of Evolution on Earth.* Princeton University Press, Princeton (2003).

Rutherford A. *Creation: The Origin of Life/The Future of Life.* Viking Press, London (2013).

Zahnle K, Arndt N, Cockell C, Halliday A, Nisbet E, Selsis F, Sleep NH. Emergence of a habitable planet. *Space Science Reviews* **129**: 35–78 (2007).

The rise of oxygen

Butterfield NJ. Oxygen, animals and oceanic ventilation: an alternative view. *Geobiology* 7: 1–7 (2009).

Canfield DE. *Oxygen: A Four Billion Year History.* Princeton University Press, Princeton (2014).

Catling DC, Glein CR, Zahnle KJ, MckayCP. Why O_2 is required by complex life on habitable planets and the concept of planetary 'oxygenation time'. *Astrobiology* **5**: 415–38 (2005).

Holland HD. The oxygenation of the atmosphere and oceans. *Philosophical Transactions Royal Society B* **361**: 903–15 (2006).

Lane N. Life's a gas. *New Scientist* **2746**: 36–39 (2010).

Lane N. *Oxygen: The Molecule that Made the World.* Oxford University Press, Oxford (2002).

Shields-Zhou G, Och L. The case for a Neoproterozoic oxygenation event: Geochemical evidence and biological consequences. *GSA Today* **21**: 4–11 (2011).

Predictions of the serial endosymbiosis hypothesis

Archibald JM. Origin of eukaryotic cells: 40 years on. *Symbiosis* **54**: 69–86 (2011).

Margulis L. Genetic and evolutionary consequences of symbiosis. *Experimental Parasitology* **39**: 277–349 (1976).

O'Malley M. The first eukaryote cell: an unfinished history of contestation. *Studies in History and Philosophy of Biological and Biomedical Sciences* 41: 212–24 (2010).

The rise and fall of Archezoa

Cavalier-Smith T. Archaebacteria and archezoa. *Nature* **339**: 100–101 (1989).

Cavalier-Smith T. Predation and eukaryotic origins: A coevolutionary perspective. *International Journal of Biochemistry and Cell Biology* **41**: 307–32 (2009).

Henze K, Martin W. Essence of mitochondria. *Nature* **426**: 127–28 (2003).

Martin WF, Müller M. *Origin of Mitochondria and Hydrogenosomes.* Springer, Heidelberg (2007).

Tielens AGM, Rotte C, Hellemond JJ, Martin W. Mitochondria as we don't know them. *Trends in Biochemical Sciences* **27**: 564–72 (2002).

van der Giezen M. Hydrogenosomes and mitosomes: Conservation and evolution of functions. *Journal of Eukaryotic Microbiology* **56**: 221–31 (2009).

Yong E. The unique merger that made you (and ewe and yew). *Nautilus* **17**: Sept 4 (2014).

Eukaryotic supergroups

Baldauf SL, Roger AJ, Wenk-Siefert I, Doolittle WF. A kingdom-level phylogeny of eukaryotes based on combined protein data. *Science* **290**: *972–77* (2000).

Hampl V, Huga L, Leigh JW, Dacks JB, Lang BF, Simpson AGB, Roger AJ. Phylogenomic analyses support the monophyly of Excavata and resolve relationships among eukaryotic 'supergroups'. *Proceedings National Academy Sciences USA* **106**: 3859–64 (2009).

Keeling PJ, Burger G, Durnford DG, Lang BF, Lee RW, Pearlman RE, Roger AJ, Grey MW. The Tree of eukaryotes. *Trends in Ecology and Evolution* **20**: 670–76 (2005).

The last eukaryotic common ancestor

Embley TM, Martin W. Eukaryotic evolution, changes and challenges. *Nature* **440**: 623–30 (2006).

Harold F. *In Search of Cell History: The Evolution of Life's Building Blocks.* Chicago University Press, Chicago (2014).

Koonin EV. The origin and early evolution of eukaryotes in the light of phylogenomics. *Genome Biology* **11**: 209 (2010).

McInerney JO, Martin WF, Koonin EV, Allen JF, Galperin MY, Lane N, Archibald JM, Embley TM. Planctomycetes and eukaryotes: a case of analogy not homology. *BioEssays* **33**: 810–17 (2011).

The paradox of small steps to complexity

Darwin C. *On the Origin of Species by Means of Natural Selection, or the Preservation of Favoured Races in the Struggle for Life* (1st Edition). John Murray, London (1859).

Land MF, Nilsson D-E. *Animal Eyes.* Oxford University Press, Oxford (2002).

Lane N. Bioenergetic constraints on the evolution of complex life. *Cold Spring Harbor Perspectives in Biology.* **doi**: 10.1101/cshperspect.a015982 (2014).

Lane N. Energetics and genetics across the prokaryote-eukaryote divide. *Biology Direct* **6**: 35 (2011).

Müller M, Mentel M, van Hellemond JJ, Henze K, Woehle C, Gould SB, Yu RY, van der Giezen M, Tielens AG, Martin WF. Biochemistry and evolution of anaerobic energy metabolism in eukaryotes. *Microbiology and Molecular Biology Reviews* **76**: 444–95 (2012).

Chapter 2: What is Living?

Energy, entropy and structure

Amend JP, LaRowe DE, McCollom TM, Shock EL. The energetics of organic synthesis inside and outside the cell. *Philosophical Transactions Royal Society B*. **368**: 20120255 (2013).

Battley EH. *Energetics of Microbial Growth*. Wiley Interscience, New York (1987).

Hansen LD, Criddle RS, Battley EH. Biological calorimetry and the thermodynamics of the origination and evolution of life. *Pure and Applied Chemistry* **81**: 1843–55 (2009).

McCollom T, Amend JP. A thermodynamic assessment of energy requirements for biomass synthesis by chemolithoautotrophic micro-organisms in oxic and micro-oxic environments. *Geobiology* **3**: 135–44 (2005).

Minsky A, Shimoni E, Frenkiel-Krispin D. Stress, order and survival. *Nature Reviews in Molecular Cell Biology* **3**: 50–60 (2002).

Rates of ATP synthesis

Fenchel T, Finlay BJ. Respiration rates in heterotrophic, free-living protozoa. *Microbial Ecology* **9**: 99–122 (1983).

Makarieva AM, Gorshkov VG, Li BL. Energetics of the smallest: do bacteria breathe at the same rate as whales? *Proceedings Royal Society B* **272**: 2219–24 (2005).

Phillips R, Kondev J, Theriot J, Garcia H. *Physical Biology of the Cell*. Garland Science, New York (2012).

Rich PR. The cost of living. *Nature* **421**: 583 (2003).

Schatz G. The tragic matter. *FEBS Letters* **536**: 1–2 (2003).

Mechanism of respiration and ATP synthesis

Abrahams JP, Leslie AG, Lutter R, Walker JE. Structure at 2.8 A resolution of F1-ATPase from bovine heart mitochondria. *Nature* **370**: 621–28 (1994).

Baradaran R, Berrisford JM, Minhas SG, Sazanov LA. Crystal structure of the entire respiratory complex I. *Nature* **494**: 443–48 (2013).

Hayashi T, Stuchebrukhov AA. Quantum electron tunneling in respiratory complex I. *Journal of Physical Chemistry B* **115**: 5354–64 (2011).

Moser CC, Page CC, Dutton PL. Darwin at the molecular scale: selection and variance in electron tunnelling proteins including cytochrome c oxidase. *Philosophical Transactions Royal Society B* **361**: 1295–1305 (2006).

Murata T, Yamato I, Kakinuma Y, Leslie AGW, Walker JE. Structure of the rotor of the V-type Na$^+$-ATPase from *Enterococcus hirae*. *Science* **308**: 654–59 (2005).

Nicholls DG, Ferguson SJ. *Bioenergetics*. Fourth Edition. Academic Press, London (2013).

Stewart AG, Sobti M, Harvey RP, Stock D. Rotary ATPases: Models, machine elements and technical specifications. *BioArchitecture* **3**: 2–12 (2013).

Vinothkumar KR, Zhu J, Hirst J. Architecture of the mammalian respiratory complex I. *Nature* **515**: 80–84 (2014).

Peter Mitchell and chemiosmotic coupling

Harold FM. *The Way of the Cell: Molecules, Organisms, and the Order of Life*. Oxford University Press, New York (2003).

Lane N. *Power, Sex, Suicide: Mitochondria and the Meaning of Life*. Oxford University Press, Oxford (2005).

Mitchell P. Coupling of phosphorylation to electron and hydrogen transfer by a chemi-osmotic type of mechanism. *Nature* **191**: 144–48 (1961).

Mitchell P. Keilin's respiratory chain concept and its chemiosmotic consequences. *Science* **206**: 1148–59 (1979).

Mitchell P. The origin of life and the formation and organising functions of natural membranes. In *Proceedings of the first international symposium on the origin of life on the Earth* (eds AI Oparin, AG Pasynski, AE Braunstein, TE Pavlovskaya). Moscow Academy of Sciences, USSR (1957).

Prebble J, Weber B. *Wandering in the Gardens of the Mind*. Oxford University Press, New York (2003).

Carbon and the need for redox chemistry

Falkowski P. *Life's Engines: How Microbes made Earth Habitable*. Princeton University Press, Princeton (2015).

Kim JD, Senn S, Harel A, Jelen BI, Falkowski PG. Discovering the electronic circuit diagram of life: structural relationships among transition metal binding sites in oxidoreductases. *Philosophical Transactions Royal Society B* **368**: 20120257 (2013).

Morton O. *Eating the Sun: How Plants Power the Planet*. Fourth Estate, London (2007).

Pace N. The universal nature of biochemistry. *Proceedings National Academy Sciences USA* **98**: 805–808 (2001).

Schoepp-Cothenet B, van Lis R, Atteia A, Baymann F, Capowiez L, Ducluzeau A-L, Duval S, ten Brink F, Russell MJ, Nitschke W. On the universal core of bioenergetics. *Biochimica Biophysica Acta Bioenergetics* **1827**: 79–93 (2013).

Fundamental differences between bacteria and archaea

Edgell DR, Doolittle WF. Archaea and the origin(s) of DNA replication proteins. *Cell* **89**: 995–98 (1997).

Koga Y, Kyuragi T, Nishihara M, Sone N. Did archaeal and bacterial cells arise independently from noncellular precursors? A hypothesis stating that the advent of membrane phospholipid with enantiomeric glycerophosphate backbones caused the separation of the two lines of descent. *Journal of Molecular Evolution* **46**: 54–63 (1998).

Leipe DD, Aravind L, Koonin EV. Did DNA replication evolve twice independently? *Nucleic Acids Research* **27**: 3389–3401 (1999).

Lombard J, López-García P, Moreira D. The early evolution of lipid membranes and the three domains of life. *Nature Reviews Microbiology* **10**: 507–15 (2012).

Martin W, Russell MJ. On the origins of cells: a hypothesis for the evolutionary transitions from abiotic geochemistry to chemoautotrophic prokaryotes, and from prokaryotes to nucleated cells. *Philosophical Transactions Royal Society B* **358**: 59–83 (2003).

Sousa FL, Thiergart T, Landan G, Nelson-Sathi S, Pereira IAC, Allen JF, Lane N, Martin WF. Early bioenergetic evolution. *Philosophical Transactions Royal Society B* **368**: 20130088 (2013).

Chapter 3: Energy at Life's Origin

Energy requirements at the origin of life

Lane N, Allen JF, Martin W. How did LUCA make a living? Chemiosmosis in the origin of life. *BioEssays* **32**: 271–80 (2010).

Lane N, Martin W. The origin of membrane bioenergetics. *Cell* **151**: 1406–16 (2012).

Martin W, Sousa FL, Lane N. Energy at life's origin. *Science* **344**: 1092–93 (2014).

Martin WF. Hydrogen, metals, bifurcating electrons, and proton gradients: The early evolution of biological energy conservation. *FEBS Letters* **586**: 485–93 (2012).

Russell M (editor). *Origins: Abiogenesis and the Search for Life*. Cosmology Science Publishers, Cambridge MA (2011).

Miller-Urey experiment and the RNA world

Joyce GF. RNA evolution and the origins of life. *Nature* **33**: 217–24 (1989).

Miller SL. A production of amino acids under possible primitive earth conditions. *Science* **117**: 528–29 (1953).

Orgel LE. Prebiotic chemistry and the origin of the RNA world. *Critical Reviews in Biochemistry and Molecular Biology* **39**: 99–123 (2004).

Powner MW, Gerland B, Sutherland JD. Synthesis of activated pyrimidine ribonucleotides in prebiotically plausible conditions. *Nature* **459**: 239–42 (2009).

Far-from-equilibrium thermodynamics

Morowitz H. *Energy Flow in Biology: Biological Organization as a Problem in Thermal Physics*. Academic Press, New York (1968).

Prigogine I. *The End of Certainty: Time, Chaos and the New Laws of Nature*. Free Press, New York (1997).

Russell MJ, Nitschke W, Branscomb E. The inevitable journey to being. *Philosophical Transactions Royal Society B* 368: 20120254 (2013).

Origins of catalysis

Cody G. Transition metal sulfides and the origins of metabolism. *Annual Review Earth and Planetary Sciences* 32: 569–99 (2004).

Russell MJ, Allen JF, Milner-White EJ. Inorganic complexes enabled the onset of life and oxygenic photosynthesis. In Allen JF, Gantt E, Golbeck JH, Osmond B: *Energy from the Sun: 14th International Congress on Photosynthesis*. Springer, Heidelberg (2008).

Russell MJ, Martin W. The rocky roots of the acetyl-CoA pathway. *Trends in Biochemical Sciences* 29: 358–63 (2004).

Dehydration reactions in water

Benner SA, Kim H-J, Carrigan MA. Asphalt, water, and the prebiotic synthesis of ribose, ribonucleosides, and RNA. *Accounts of Chemical Research* 45: 2025–34 (2012).

de Zwart II, Meade SJ, Pratt AJ. Biomimetic phosphoryl transfer catalysed by iron(II)-mineral precipitates. *Geochimica et Cosmochimica Acta* 68: 4093–98 (2004).

Pratt AJ. Prebiological evolution and the metabolic origins of life. *Artificial Life* 17: 203–17 (2011).

Formation of protocells

Budin I, Bruckner RJ, Szostak JW. Formation of protocell-like vesicles in a thermal diffusion column. *Journal of the American Chemical Society* 131: 9628–29 (2009).

Errington J. L-form bacteria, cell walls and the origins of life. *Open Biology* 3: 120143 (2013).

Hanczyc M, Fujikawa S, Szostak J. Experimental models of primitive cellular compartments: encapsulation, growth, and division. *Science* 302: 618–22 (2003).

Mauer SE, Monndard PA. Primitive membrane formation, characteristics and roles in the emergent properties of a protocell. *Entropy* 13: 466–84 (2011).

Szathmáry E, Santos M, Fernando C. Evolutionary potential and requirements for minimal protocells. *Topics in Current Chemistry* 259: 167–211 (2005).

Origins of replication

Cairns-Smith G. *Seven Clues to the Origin of Life*. Cambridge University Press, Cambridge (1990).

Costanzo G, Pino S, Ciciriello F, Di Mauro E. Generation of long RNA chains in water. *Journal of Biological Chemistry* 284: 33206–16 (2009).

Koonin EV, Martin W. On the origin of genomes and cells within inorganic compartments. *Trends in Genetics* 21: 647–54 (2005).

Mast CB, Schink S, Gerland U & Braun D. Escalation of polymerization in a thermal gradient. *Proceedings of the National Academy of Sciences USA* 110: 8030–35 (2013).

Mills DR, Peterson RL, Spiegelman S. An extracellular Darwinian experiment with a self-duplicating nucleic acid molecule. *Proceedings National Academy Sciences USA* 58: 217–24 (1967).

Discovery of deep-sea hydrothermal vents

Baross JA, Hoffman SE. Submarine hydrothermal vents and associated gradient environments as sites for the origin and evolution of life. *Origins Life Evolution of the Biosphere* 15: 327–45 (1985).

Kelley DS, Karson JA, Blackman DK, *et al.* An off-axis hydrothermal vent field near the Mid-Atlantic Ridge at 30 degrees N. *Nature* 412: 145–49 (2001).

Kelley DS, Karson JA, Früh-Green GL, *et al.* A serpentinite-hosted submarine ecosystem: the Lost City Hydrothermal Field. *Science* 307: 1428–34 (2005).

Pyrites pulling and the iron-sulphur world

de Duve C, Miller S. Two-dimensional life? *Proceedings National Academy Sciences USA* 88: 10014–17 (1991).

Huber C, Wächtershäuser G. Activated acetic acid by carbon fixation on (Fe,Ni)S under primordial conditions. *Science* 276: 245–47 (1997).

Miller SL, Bada JL. Submarine hot springs and the origin of life. *Nature* 334: 609–611 (1988).

Wächtershäuser G. Evolution of the first metabolic cycles. *Proceedings National Academy Sciences USA* 87: 200–204 (1990).

Wächtershäuser G. From volcanic origins of chemoautotrophic life to Bacteria, Archaea and Eukarya. *Philosophical Transactions Royal Society B* 361: 1787–1806 (2006).

Alkaline hydrothermal vents

Martin W, Baross J, Kelley D, Russell MJ. Hydrothermal vents and the origin of life. *Nature Reviews Microbiology* 6: 805–14 (2008).

Martin W, Russell MJ. On the origins of cells: a hypothesis for the evolutionary transitions from abiotic geochemistry to chemoautotrophic prokaryotes, and from prokaryotes to nucleated cells. *Philosophical Transactions Royal Society B* 358: 59–83 (2003).

Russell MJ, Daniel RM, Hall AJ, Sherringham J. A hydrothermally precipitated catalytic iron sulphide membrane as a first step toward life. *Journal of Molecular Evolution* **39**: 231–43 (1994).

Russell MJ, Hall AJ, Cairns-Smith AG, Braterman PS. Submarine hot springs and the origin of life. *Nature* **336**: 117 (1988).

Russell MJ, Hall AJ. The emergence of life from iron monosulphide bubbles at a submarine hydrothermal redox and pH front. *Journal Geological Society London* **154**: 377–402 (1997).

Serpentinization

Fyfe WS. The water inventory of the Earth: fluids and tectonics. *Geological Society of London Special Publications* **78**: 1–7 (1994).

Russell MJ, Hall AJ, Martin W. Serpentinization as a source of energy at the origin of life. *Geobiology* **8**: 355–71 (2010).

Sleep NH, Bird DK, Pope EC. Serpentinite and the dawn of life. *Philosophical Transactions Royal Society B* **366**: 2857–69 (2011).

Ocean chemistry in the Hadean

Arndt N, Nisbet E. Processes on the young earth and the habitats of early life. *Annual Reviews Earth Planetary Sciences* **40**: 521–49 (2012).

Pinti D. The origin and evolution of the oceans. *Lectures Astrobiology* **1**: 83–112 (2005).

Russell MJ, Arndt NT. Geodynamic and metabolic cycles in the Hadean. *Biogeosciences* **2**: 97–111 (2005).

Zahnle K, Arndt N, Cockell C, Halliday A, Nisbet E, Selsis F, Sleep NH. Emergence of a habitable planet. *Space Science Reviews* **129**: 35–78 (2007).

Thermophoresis

Baaske P, Weinert FM, Duhr S, et al. Extreme accumulation of nucleotides in simulated hydrothermal pore systems. *Proceedings National Academy Sciences USA* **104**: 9346–51 (2007).

Mast CB, Schink S, Gerland U, Braun D. Escalation of polymerization in a thermal gradient. *Proceedings National Academy Sciences USA* **110**: 8030–35 (2013).

Thermodynamics of organic synthesis in alkaline vents

Amend JP, McCollom TM. Energetics of biomolecule synthesis on early Earth. In Zaikowski L *et al.* eds. *Chemical Evolution II: From the Origins of Life to Modern Society*. American Chemical Society (2009).

Ducluzeau A-L, Schoepp-Cothenet B, Baymann F, Russell MJ, Nitschke W. Free energy conversion in the LUCA: Quo vadis? *Biochimica et Biophysica Acta Bioenergetics* **1837**: 982–988 (2014).

Martin W, Russell MJ. On the origin of biochemistry at an alkaline hydrothermal vent. *Philosophical Transactions Royal Society B* **367**: 1887–1925 (2007).

Shock E, Canovas P. The potential for abiotic organic synthesis and biosynthesis at seafloor hydrothermal systems. *Geofluids* **10**: 161–92 (2010).

Sousa FL, Thiergart T, Landan G, Nelson-Sathi S, Pereira IAC, Allen JF, Lane N, Martin WF. Early bioenergetic evolution. *Philosophical Transactions Royal Society B* **368**: 20130088 (2013).

Reduction potential and the kinetic barrier to CO_2 reduction

Lane N, Martin W. The origin of membrane bioenergetics. *Cell* **151**: 1406–16 (2012).

Maden BEH. Tetrahydrofolate and tetrahydromethanopterin compared: functionally distinct carriers in C1 metabolism. *Biochemical Journal* **350**: 609–29 (2000).

Wächtershäuser G. Pyrite formation, the first energy source for life: a hypothesis. *Systematic and Applied Microbiology* **10**: 207–10 (1988).

Could natural proton gradients drive the reduction of CO_2?

Herschy B, Whicher A, Camprubi E, Watson C, Dartnell L, Ward J, Evans JRG, Lane N. An origin-of-life reactor to simulate alkaline hydrothermal vents. *Journal of Molecular Evolution* **79**: 213–27 (2014).

Herschy B. Nature's electrochemical flow reactors: Alkaline hydrothermal vents and the origins of life. *Biochemist* **36**: 4–8 (2014).

Lane N. Bioenergetic constraints on the evolution of complex life. *Cold Spring Harbor Perspectives in Biology* **doi**: 10.1101/cshperspect.a015982 (2014).

Nitschke W, Russell MJ. Hydrothermal focusing of chemical and chemiosmotic energy, supported by delivery of catalytic Fe, Ni, Mo, Co, S and Se forced life to emerge. *Journal of Molecular Evolution* **69**: 481–96 (2009).

Yamaguchi A, Yamamoto M, Takai K, Ishii T, Hashimoto K, Nakamura R. Electrochemical CO_2 reduction by Nicontaining iron sulfides: how is Co2 electrochemically reduced at bisulfide-bearing deep sea hydrothermal precipitates? *Electrochimica Acta* **141**: 311–18 (2014).

Probability of serpentinization in the Milky Way

de Leeuw NH, Catlow CR, King HE, Putnis A, Muralidharan K, Deymier P, Stimpfl M, Drake MJ. Where on Earth has our water come from? *Chemical Communications* **46**: 8923–25 (2010).

Petigura EA, Howard AW, Marcy GW. Prevalence of Earth-sized planets orbiting Sun-like stars. *Proceedings National Academy Sciences USA* **110**: 19273–78 (2013).

Chapter 4: The Emergence of Cells

The problem of lateral gene transfer and speciation

Doolittle WF. Phylogenetic classification and the universal tree. *Science* **284**: 2124–28 (1999).

Lawton G. Why Darwin was wrong about the tree of life. *New Scientist* **2692**: 34–39 (2009).

Mallet J. Why was Darwin's view of species rejected by twentieth century biologists? *Biology and Philosophy* **25**: 497–527 (2010).

Martin WF. Early evolution without a tree of life. *Biology Direct* **6**: 36 (2011).

Nelson-Sathi S *et al*. Origins of major archaeal clades correspond to gene acquisitions from bacteria. *Nature* **doi**: 10.1038/nature13805 (2014).

The 'universal tree of life' based on fewer than 1% of genes

Ciccarelli FD, Doerks T, von Mering C, Creevey CJ, Snel B, *et al*. Toward automatic reconstruction of a highly resolved tree of life. *Science* **311**: 1283–87 (2006).

Dagan T, Martin W. The tree of one percent. *Genome Biology* **7**: 118 (2006).

Genes conserved in archaea and bacteria

Charlebois RL, Doolittle WF. Computing prokaryotic gene ubiquity: Rescuing the core from extinction. *Genome Research* **14**: 2469–77 (2004).

Koonin EV. Comparative genomics, minimal gene-sets and the last universal common ancestor. *Nature Reviews Microbiology* **1**: 127–36 (2003).

Sousa FL, Thiergart T, Landan G, Nelson-Sathi S, Pereira IAC, Allen JF, Lane N, Martin WF. Early bioenergetic evolution. *Philosophical Transactions of the Royal Society B* **368**: 20130088 (2013).

Paradoxical properties of LUCA

Dagan T, Martin W. Ancestral genome sizes specify the minimum rate of lateral gene transfer during prokaryote evolution. *Proceedings National Academy Sciences USA* **104**: 870–75 (2007).

Edgell DR, Doolittle WF. Archaea and the origin(s) of DNA replication proteins. *Cell* **89**: 995–98 (1997).

Koga Y, Kyuragi T, Nishihara M, Sone N. Did archaeal and bacterial cells arise independently from noncellular precursors? A hypothesis stating that the advent of membrane phospholipid with enantiomeric glycerophosphate backbones caused the separation of the two lines of descent. *Journal of Molecular Evolution* **46**: 54–63 (1998).

Leipe DD, Aravind L, Koonin EV. Did DNA replication evolve twice independently? *Nucleic Acids Research* **27**: 3389–3401 (1999).

Martin W, Russell MJ. On the origins of cells: a hypothesis for the evolutionary transitions from abiotic geochemistry to chemoautotrophic prokaryotes, and from prokaryotes to nucleated cells. *Philosophical Transactions Royal Society B* **358**: 59–83 (2003).

The problem of membrane lipids

Lane N, Martin W. The origin of membrane bioenergetics. *Cell* 151: 1406–16 (2012).

Lombard J, López-García P, Moreira D. The early evolution of lipid membranes and the three domains of life. *Nature Reviews in Microbiology* **10**: 507–15 (2012).

Shimada H, Yamagishi A. Stability of heterochiral hybrid membrane made of bacterial sn-G3P lipids and archaeal sn-G1P lipids. *Biochemistry* **50**: 4114–20 (2011).

Valentine D. Adaptations to energy stress dictate the ecology and evolution of the Archaea. *Nature Reviews Microbiology* **5**: 1070–77 (2007).

Acetyl CoA pathway

Fuchs G. Alternative pathways of carbon dioxide fixation: Insights into the early evolution of life? *Annual Review Microbiology* **65**: 631–58 (2011).

Ljungdahl LG. A life with acetogens, thermophiles, and cellulolytic anaerobes . *Annual Review Microbiology* **63**: 1–25 (2009).

Maden BEH. No soup for starters? Autotrophy and the origins of metabolism. *Trends in Biochemical Sciences* **20**: 337–41 (1995).

Ragsdale SW, Pierce E. Acetogenesis and the Wood-Ljungdahl pathway of CO_2 fixation. *Biochimica Biophysica Acta* **1784**: 1873–98 (2008).

Rocky roots of the acetyl CoA pathway

Nitschke W, McGlynn SE, Milner-White J, Russell MJ. On the antiquity of metalloenzymes and their substrates in bioenergetics. *Biochimica Biophysica Acta* **1827**: 871–81 (2013).

Russell MJ, Martin W. The rocky roots of the acetyl-CoA pathway. *Trends in Biochemical Sciences* **29**: 358–63 (2004).

Abiotic synthesis of acetyl thioesters and acetyl phosphate

de Duve C. Did God make RNA? *Nature* **336**: 209–10 (1988).

Heinen W, Lauwers AM. Sulfur compounds resulting from the interaction of iron sulfide, hydrogen sulfide and carbon dioxide in an anaerobic aqueous environment. *Origins Life Evolution Biosphere* **26**: 131–50 (1996).

Huber C, Wächtershäuser G. Activated acetic acid by carbon fixation on (Fe,Ni)S under primordial conditions. *Science* **276**: 245–47 (1997).

Martin W, Russell MJ. On the origin of biochemistry at an alkaline hydrothermal vent. *Philosophical Transactions of the Royal Society B* **367**: 1887–1925 (2007).

Possible origins of the genetic code

Copley SD, Smith E, Morowitz HJ. A mechanism for the association of amino acids with their codons and the origin of the genetic code. *Proceedings National Academy Sciences USA* **102**: 4442–47 (2005).

Lane N. *Life Ascending: The Ten Great Inventions of Evolution.* WW Norton/Profile, London (2009).

Taylor FJ. Coates D. The code within the codons. *Biosystems* **22**: 177–87 (1989).

Congruence between alkaline hydrothermal vents and acetyl CoA pathway

Herschy B, Whicher A, Camprubi E, Watson C, Dartnell L, Ward J, Evans JRG, Lane N. An origin-of-life reactor to simulate alkaline hydrothermal vents. *Journal of Molecular Evolution* **79**: 213–27 (2014).

Lane N. Bioenergetic constraints on the evolution of complex life. *Cold Spring Harbor Perspectives in Biology* doi: 10.1101/cshperspect.a015982 (2014).

Martin W, Sousa FL, Lane N. Energy at life's origin. *Science* **344**: 1092–93 (2014).

Sousa FL, Thiergart T, Landan G, Nelson-Sathi S, Pereira IAC, Allen JF, Lane N, Martin WF. Early bioenergetic evolution. *Philosophical Transactions of the Royal Society B* **368**: 20130088 (2013).

The problem of membrane permeability

Lane N, Martin W. The origin of membrane bioenergetics. *Cell* **151**: 1406–16 (2012).

Le Page M. Meet your maker. *New Scientist* **2982**: 30–33 (2014).

Mulkidjanian AY, Bychkov AY, Dibrova D V, Galperin MY, Koonin EV. Origin of first cells at terrestrial, anoxic geothermal fields. *Proceedings National Academy Sciences USA* **109**: E821–E830 (2012).

Sojo V, Pomiankowski A, Lane N. A bioenergetic basis for membrane divergence in archaea and bacteria. *PLoS Biology* **12(8)**: e1001926 (2014).

Yong E. How life emerged from deep-sea rocks. *Nature* doi: 10.1038/nature.2012.12109 (2012).

Promiscuity of membrane proteins for H⁺ and Na⁺

Buckel W, Thauer RK. Energy conservation via electron bifurcating ferredoxin reduction and proton/Na(+) translocating ferredoxin oxidation. *Biochimica Biophysica Acta* **1827**: 94–113 (2013).

Lane N, Allen JF, Martin W. How did LUCA make a living? Chemiosmosis in the origin of life. *BioEssays* **32**: 271–80 (2010).

Schlegel K, Leone V, Faraldo-Gómez JD, Müller V. Promiscuous archaeal ATP synthase concurrently coupled to Na⁺ and H⁺ translocation. *Proceedings National Academy Sciences USA* **109**: 947–52 (2012).

Electron bifurcation

Buckel W, Thauer RK. Energy conservation via electron bifurcating ferredoxin reduction and proton/Na(+) translocating ferredoxin oxidation. *Biochimica Biophysica Acta* **1827**: 94–113 (2013).

Kaster A-K, Moll J, Parey K, Thauer RK. Coupling of ferredoxin and heterodisulfide reduction via electron bifurcation in hydrogenotrophic methanogenic Archaea. *Proceedings National Academy Sciences USA* **108**: 2981–86 (2011).

Thauer RK. A novel mechanism of energetic coupling in anaerobes. *Environmental Microbiology Reports* **3**: 24–25 (2011).

Chapter 5: The Origin of Complex Cells

Genome sizes

Cavalier-Smith T. Economy, speed and size matter: evolutionary forces driving nuclear genome miniaturization and expansion. *Annals of Botany* **95**: 147–75 (2005).

Cavalier-Smith T. Skeletal DNA and the evolution of genome size. *Annual Review of Biophysics and Bioengineering* **11**: 273–301 (1982).

Gregory TR. Synergy between sequence and size in large-scale genomics. *Nature Reviews in Genetics* **6**: 699–708 (2005).

Lynch M. *The Origins of Genome Architecture*. Sinauer Associates, Sunderland MA (2007).

Possible constraints on eukaryotic genome size

Cavalier-Smith T. Predation and eukaryote cell origins: A coevolutionary perspective. *International Journal Biochemistry Cell Biology* **41**: 307–22 (2009).

de Duve C. The origin of eukaryotes: a reappraisal. *Nature Reviews in Genetics* **8**: 395–403 (2007).

Koonin EV. Evolution of genome architecture. *International Journal Biochemistry Cell Biology* **41**: 298–306 (2009).

Lynch M, Conery JS. The origins of genome complexity. *Science* **302**: 1401–04 (2003).

Maynard Smith J, Szathmary E. *The Major Transitions in Evolution.* Oxford University Press, Oxford. (1995).

Chimeric origin of eukaryotes

Cotton JA, McInerney JO. Eukaryotic genes of archaebacterial origin are more important than the more numerous eubacterial genes, irrespective of function. *Proceedings National Academy Sciences USA* **107**: 17252–55 (2010).

Esser C, Ahmadinejad N, Wiegand C, et al. A genome phylogeny for mitochondria among alpha-proteobacteria and a predominantly eubacterial ancestry of yeast nuclear genes. *Molecular Biology Evolution* **21**: 1643–60 (2004).

Koonin EV. Darwinian evolution in the light of genomics. *Nucleic Acids Research* **37**: 1011–34 (2009).

Pisani D, Cotton JA, McInerney JO. Supertrees disentangle the chimeric origin of eukaryotic genomes. *Molecular Biology Evolution* **24**: 1752–60 (2007).

Rivera MC, Lake JA. The ring of life provides evidence for a genome fusion origin of eukaryotes. *Nature* **431**: 152–55 (2004).

Thiergart T, Landan G, Schrenk M, Dagan T, Martin WF. An evolutionary network of genes present in the eukaryote common ancestor polls genomes on eukaryotic and mitochondrial origin. *Genome Biology and Evolution* **4**: 466–85 (2012).

Williams TA, Foster PG, Cox CJ, Embley TM. An archaeal origin of eukaryotes supports only two primary domains of life. *Nature* **504**: 231–36 (2013).

Late origin of fermentation

Say RF, Fuchs G. Fructose 1,6-bisphosphate aldolase/phosphatase may be an ancestral gluconeogenic enzyme. *Nature* **464**: 1077–81 (2010).

Sub-stochiometric energy conservation

Hoehler TM, Jørgensen BB. Microbial life under extreme energy limitation. *Nature Reviews in Microbiology* **11**: 83–94 (2013).

Lane N. Why are cells powered by proton gradients? *Nature Education* **3**: 18 (2010).

Martin W, Russell MJ. On the origin of biochemistry at an alkaline hydrothermal vent. *Philosophical Transactions of the Royal Society B* **367**: 1887–1925 (2007).

Thauer RK, Kaster A-K, Seedorf H, Buckel W, Hedderich R. Methanogenic archaea: ecologically relevant differences in energy conservation. *Nature Reviews Microbiology* **6**: 579–91 (2007).

Viral infection and cell death

Bidle KD, Falkowski PG. Cell death in planktonic, photosynthetic microorganisms. *Nature Reviews Microbiology* **2**: 643–55 (2004).

Lane N. Origins of death. *Nature* **453**: 583–85 (2008).

Refardt D, Bergmiller T, Kümmerli R. Altruism can evolve when relatedness is low: evidence from bacteria committing suicide upon phage infection. *Proceedings Royal Society B* **280**: 20123035 (2013).

Vardi A, Formiggini F, Casotti R, De Martino A, Ribalet F, Miralto A, Bowler C. A stress surveillance system based on calcium and nitroc oxide in marine diatoms. *PLoS Biology* **4(3)**: e60 (2006).

Scaling of bacterial surface area and volume

Fenchel T, Finlay BJ. Respiration rates in heterotrophic, free-living protozoa. *Microbial Ecology* **9**: 99–122 (1983).

Harold F. *The Vital Force: a Study of Bioenergetics.* WH Freeman, New York (1986).

Lane N, Martin W. The energetics of genome complexity. *Nature* **467**: 929–34 (2010).

Lane N. Energetics and genetics across the prokaryote-eukaryote divide. *Biology Direct* **6**: 35 (2011).

Makarieva AM, Gorshkov VG, Li BL. Energetics of the smallest: do bacteria breathe at the same rate as whales? *Proceedings Royal Society B* **272**: 2219–24 (2005).

Vellai T, Vida G. The origin of eukaryotes: the difference between prokaryotic and eukaryotic cells. *Proceedings Royal Society B* **266**: 1571–77 (1999).

Giant bacteria

Angert ER. DNA replication and genomic architecture of very large bacteria. *Annual Review Microbiology* **66**: 197–212 (2012).

Mendell JE, Clements KD, Choat JH, Angert ER. Extreme polyploidy in a large bacterium. *Proceedings National Academy Sciences USA* **105**: 6730–34 (2008).

Schulz HN, Jorgensen BB. Big bacteria. *Annual Review Microbiology* **55**: 105–37 (2001).

Schulz HN. The genus *Thiomargarita*. *Prokaryotes* **6**: 1156–63 (2006).

Small endosymbiont genomes and energetic consequences

Gregory TR, DeSalle R. Comparative genomics in prokaryotes. In *The Evolution of the Genome* ed. Gregory TR. Elsevier, San Diego, pp. 585–75 (2005).

Lane N, Martin W. The energetics of genome complexity. *Nature* **467**: 929–34 (2010).

Lane N. Bioenergetic constraints on the evolution of complex life. *Cold Spring Harbor Perspectives in Biology* **doi:** 10.1101/cshperspect.a015982 (2014).

Endosymbionts in bacteria

von Dohlen CD, Kohler S, Alsop ST, McManus WR. Mealybug beta-proteobacterial symbionts contain gamma-proteobacterial symbionts. *Nature* **412**: 433–36 (2001).

Wujek DE. Intracellular bacteria in the blue-green-alga *Pleurocapsa minor*. *Transactions American Microscopical Society* **98**: 143–45 (1979).

Why mitochondria retain genes

Alberts A, Johnson A, Lewis J, Raff M, Roberts K, Walter P. *Molecular Biology of the Cell*, 5th edition. Garland Science, New York (2008).

Allen JF. Control of gene expression by redox potential and the requirement for chloroplast and mitochondrial genomes. *Journal of Theoretical Biology* **165**: 609–31 (1993).

Allen JF. The function of genomes in bioenergetic organelles. *Philosophical Transactions Royal Society B* **358**: 19–37 (2003).

de Grey AD. Forces maintaining organellar genomes: is any as strong as genetic code disparity or hydrophobicity? *BioEssays* **27**: 436–46 (2005).

Gray MW, Burger G, Lang BF. Mitochondrial evolution. *Science* **283**: 1476–81 (1999).

Polyploidy in cyanobacteria

Griese M, Lange C, Soppa J. Ploidy in cyanobacteria. *FEMS Microbiology Letters* **323**: 124–31 (2011).

Why plastids can't overcome the energetic constraints on bacteria

Lane N. Bioenergetic constraints on the evolution of complex life. *Cold Spring Harbor Perspectives in Biology* **doi**: 10.1101/cshperspect.a015982 (2014).

Lane N. Energetics and genetics across the prokaryote-eukaryote divide. *Biology Direct* **6**: 35 (2011).

Levels of selection conflict and resolution in endosymbioses

Blackstone NW. Why did eukaryotes evolve only once? Genetic and energetic aspects of conflict and conflict mediation. *Philosophical Transactions Royal Society B* **368**: 20120266 (2013).

Martin W, Müller M. The hydrogen hypothesis for the first eukaryote. *Nature* **392**: 37–41 (1998).

Energy spilling in bacteria

Russell JB. The energy spilling reactions of bacteria and other organisms. *Journal of Molecular Microbiology and Biotechnology* **13**: 1–11 (2007).

Chapter 6: Sex and the Origins of Death

Speed of evolution

Conway-Morris S. The Cambrian "explosion": Slow-fuse or megatonnage? *Proceedings National Academy Sciences USA* 97: 4426–29 (2000).

Gould SJ, Eldredge N. Punctuated equilibria: the tempo and mode of evolution reconsidered. *Paleobiology* 3: 115–51 (1977).

Nilsson D-E, Pelger S. A pessimistic estimate of the time required for an eye to evolve. *Proceedings Royal Society B* 256: 53–58 (1994).

Sex and population structure

Lahr DJ, Parfrey LW, Mitchell EA, Katz LA, Lara E. The chastity of amoeba: re-evaluating evidence for sex in amoeboid organisms. *Proceedings Royal Society B* 278: 2081–90 (2011).

Maynard-Smith J. *The Evolution of Sex.* Cambridge University Press, Cambridge (1978).

Ramesh MA, Malik SB, Logsdon JM. A phylogenomic inventory of meiotic genes: evidence for sex in *Giardia* and an early eukaryotic origin of meiosis. *Current Biology* 15: 185–91 (2005).

Takeuchi N, Kaneko K, Koonin EV. Horizontal gene transfer can rescue prokaryotes from Muller's ratchet: benefit of DNA from dead cells and population subdivision. *Genes Genomes Genetics* 4: 325–39 (2014).

The origin of introns

Cavalier-Smith T. Intron phylogeny: A new hypothesis. *Trends in Genetics* 7: 145–48 (1991).

Doolittle WF. Genes in pieces: were they ever together? *Nature* 272: 581–82 (1978).

Koonin EV. The origin of introns and their role in eukaryogenesis: a compromise solution to the introns-early versus introns-late debate? *Biology Direct* 1: 22 (2006).

Lambowitz AM, Zimmerly S. Group II introns: mobile ribozymes that invade DNA. *Cold Spring Harbor Perspectives in Biology* 3: a003616 (2011).

Introns and the origin of the nucleus

Koonin E. Intron-dominated genomes of early ancestors of eukaryotes. *Journal of Heredity* 100: 618–23 (2009).

Martin W, Koonin EV. Introns and the origin of nucleus–cytosol compartmentalization. *Nature* 440: 41–45 (2006).

Rogozin IB, Wokf YI, Sorokin AV, Mirkin BG, Koonin EV. Remarkable interkingdom conservation of intron positions and massive, lineage-specific intron loss and gain in eukaryotic evolution. *Current Biology* 13: 1512–17 (2003).

Sverdlov AV, Csuros M, Rogozin IB, Koonin EV. A glimpse of a putative pre-intron phase of eukaryotic evolution. *Trends in Genetics* **23**: 105–08 (2007).

Numts

Hazkani-Covo E, Zeller RM, Martin W. Molecular poltergeists: mitochondrial DNA copies (numts) in sequenced nuclear genomes. *PLoS Genetics* **6**: e1000834 (2010).

Lane N. Plastids, genomes and the probability of gene transfer. *Genome Biology and Evolution* **3**: 372–74 (2011).

Strength of selection against introns

Lane N. Energetics and genetics across the prokaryote-eukaryote divide. *Biology Direct* **6**: 35 (2011).

Lynch M, Richardson AO. The evolution of spliceosomal introns. *Current Opinion in Genetics and Development* **12**: 701–10 (2002).

Speed of splicing versus translation

Cavalier-Smith T. Intron phylogeny: A new hypothesis. *Trends in Genetics* **7**: 145–48 (1991).

Martin W, Koonin EV. Introns and the origin of nucleus–cytosol compartmentalization. *Nature* **440**: 41–45 (2006).

Origin of nuclear membrane, pore complexes and nucleolus

Mans BJ, Anantharaman V, Aravind L, Koonin EV. Comparative genomics, evolution and origins of the nuclear envelope and nuclear pore complex. *Cell Cycle* **3**: 1612–37 (2004).

Martin W. A briefly argued case that mitochondria and plastids are descendants of endosymbionts, but that the nuclear compartment is not. *Proceedings of the Royal Society B* **266**: 1387–95 (1999).

Martin W. Archaebacteria (Archaea) and the origin of the eukaryotic nucleus. Current Opinion in microbiology 8: 630–37 (2005).

McInerney JO, Martin WF, Koonin EV, Allen JF, Galperin MY, Lane N, Archibald JM, Embley TM. Planctomycetes and eukaryotes: A case of analogy not homology. *BioEssays* **33**: 810–17 (2011).

Mercier R, Kawai Y, Errington J. Excess membrane synthesis drives a primitive mode of cell proliferation. Cell 152: 997–1007 (2013).

Staub E, Fiziev P, Rosenthal A, Hinzmann B. Insights into the evolution of the nucleolus by an analysis of its protein domain repertoire. *BioEssays* **26**: 567–81 (2004)

The evolution of sex

Bell G. *The Masterpiece of Nature: The Evolution and Genetics of Sexuality*. University of California Press, Berkeley (1982).

Felsenstein J. The evolutionary advantage of recombination. *Genetics* 78: 737–56 (1974).

Hamilton WD. Sex versus non-sex versus parasite. *Oikos* 35: 282–90 (1980).

Lane N. Why sex is worth losing your head for. *New Scientist* 2712: 40–43 (2009).

Otto SP, Barton N. Selection for recombination in small populations. *Evolution* 55: 1921–31 (2001).

Partridge L, Hurst LD. Sex and conflict. *Science* 281: 2003–08 (1998).

Ridley M. *Mendel's Demon: Gene Justice and the Complexity of Life*. Weidenfeld and Nicholson, London (2000).

Ridley M. *The Red Queen: Sex and the Evolution of Human Nature*. Penguin, London (1994).

Possible origins of cell fusion and chromosomal segregation

Blackstone NW, Green DR. The evolution of a mechanism of cell suicide. *BioEssays* 21: 84–88 (1999).

Ebersbach G, Gerdes K. Plasmid segregation mechanisms. *Annual Review Genetics* 39: 453–79 (2005).

Errington J. L-form bacteria, cell walls and the origins of life. *Open Biology* 3: 120143 (2013).

Two sexes

Fisher RA. *The Genetical Theory of Natural Selection*. Clarendon Press, Oxford (1930).

Hoekstra RF. On the asymmetry of sex – evolution of mating types in isogamous populations. *Journal of Theoretical Biology* 98: 427–51 (1982).

Hurst LD, Hamilton WD. Cytoplasmic fusion and the nature of sexes. *Proceedings of the Royal Society B* 247: 189–94 (1992).

Hutson V, Law R. Four steps to two sexes. *Proceedings Royal Society B* 253: 43–51 (1993).

Parker GA, Smith VGF, Baker RR. The origin and evolution of gamete dimorphism and the male-female phenomenon. *Journal of Theoretical Biology* 36: 529–53 (1972).

Uniparental inheritance of mitochondria

Birky CW. Uniparental inheritance of mitochondrial and chloroplast genes – mechanisms and evolution. *Proceedings National Academy Sciences USA* 92: 11331–38 (1995).

Cosmides LM, Tooby J. Cytoplasmic inheritance and intragenomic conflict. *Journal of Theoretical Biology* 89: 83–129 (1981).

Hadjivasiliou Z, Lane N, Seymour R, Pomiankowski A. Dynamics of mitochondrial inheritance in the evolution of binary mating types and two sexes. *Proceedings Royal Society B* **280**: 20131920 (2013).

Hadjivasiliou Z, Pomiankowski A, Seymour R, Lane N. Selection for mitonuclear co-adaptation could favour the evolution of two sexes. *Proceedings Royal Society B* **279**: 1865–72 (2012).

Lane N. *Power, Sex, Suicide: Mitochondria and the Meaning of Life*. Oxford University Press, Oxford (2005).

Mitochondrial mutation rates in animals, plants and basal metazoans

Galtier N. The intriguing evolutionary dynamics of plant mitochondrial DNA. *BMC Biology* **9**: 61 (2011).

Huang D, Meier R, Todd PA, Chou LM. Slow mitochondrial *COI* sequence evolution at the base of the metazoan tree and its implications for DNA barcoding. *Journal of Molecular Evolution* **66**: 167–74 (2008).

Lane N. On the origin of barcodes. *Nature* **462**: 272–74 (2009).

Linnane AW, Ozawa T, Marzuki S, Tanaka M. *Lancet* **333**: 642–45 (1989).

Pesole G, Gissi C, De Chirico A, Saccone C. Nucleotide substitution rate of mammalian mitochondrial genomes. *Journal of Molecular Evolution* **48**: 427–34 (1999).

Origin of the germline-soma distinction

Allen JF, de Paula WBM. Mitochondrial genome function and maternal inheritance. *Biochemical Society Transactions* **41**: 1298–1304 (2013).

Allen JF. Separate sexes and the mitochondrial theory of ageing. *Journal of Theoretical Biology* **180**: 135–40 (1996).

Buss L. *The Evolution of Individuality*. Princeton University Press, Princeton (1987).

Clark WR. *Sex and the Origins of Death*. Oxford University Press, New York (1997).

Radzvilavicius AL, Hadjivasiliou Z, Pomiankowski A, Lane N. Mitochondrial variation drives the evolution of sexes and the germline-soma distinction. MS in preparation (2015).

Chapter 7: The Power and the Glory

The mosaic respiratory chain

Allen JF. The function of genomes in bioenergetic organelles. *Philosophical Transactions Royal Society B* **358**: 19–37 (2003).

Lane N. The costs of breathing. *Science* **334**: 184–85 (2011).

Moser CC, Page CC, Dutton PL. Darwin at the molecular scale: selection and variance in electron tunnelling proteins including cytochrome c oxidase. *Philosophical Transactions Royal Society B* **361**: 1295–1305 (2006).

Schatz G, Mason TL. The biosynthesis of mitochondrial proteins. *Annual Review Biochemistry* **43**: 51–87 (1974).

Vinothkumar KR, Zhu J, Hirst J. Architecture of the mammalian respiratory complex I. *Nature* **515**: 80–84 (2014).

Hybrid breakdown, cybrids and the origin of species

Barrientos A, Kenyon L, Moraes CT. Human xenomitochondrial cybrids. Cellular models of mitochondrial complex I deficiency. *Journal of Biological Chemistry* **273**: 14210–17 (1998).

Blier PU, Dufresne F, Burton RS. Natural selection and the evolution of mtDNA-encoded peptides: evidence for intergenomic co-adaptation. *Trends in Genetics* **17**: 400–406 (2001).

Burton RS, Barreto FS. A disproportionate role for mtDNA in Dobzhansky-Muller incompatibilities? *Molecular Ecology* **21**: 4942–57 (2012).

Burton RS, Ellison CK, Harrison JS. The sorry state of F2 hybrids: consequences of rapid mitochondrial DNA evolution in allopatric populations. *American Naturalist* **168** Supplement 6: S14–24 (2006).

Gershoni M, Templeton AR, Mishmar D. Mitochondrial biogenesis as a major motive force of speciation. *Bioessays* **31**: 642–50 (2009).

Lane N. On the origin of barcodes. *Nature* **462**: 272–74 (2009).

Mitochondrial control of apoptosis

Hengartner MO. Death cycle and Swiss army knives. *Nature* **391**: 441–42 (1998).

Koonin EV, Aravind L. Origin and evolution of eukaryotic apoptosis: the bacterial connection. *Cell Death and Differentiation* **9**: 394–404 (2002).

Lane N. Origins of death. *Nature* **453**: 583–85 (2008).

Zamzami N, Kroemer G. The mitochondrion in apoptosis: how pandora's box opens. *Nature Reviews Molecular Cell Biology* **2**: 67–71 (2001).

Rapid evolution of animal mitochondrial genes and environmental adaptation

Bazin E, Glémin S, Galtier N. Population size dies not influence mitochondrial genetic diversity in animals. *Science* **312**: 570–72 (2006).

Lane N. On the origin of barcodes. *Nature* **462**: 272–74 (2009).

Nabholz B, Glémin S, Galtier N. The erratic mitochondrial clock: variations of mutation rate, not population size, affect mtDNA diversity across birds and mammals. *BMC Evolutionary Biology* **9**: 54 (2009).

Wallace DC. Bioenergetics in human evolution and disease: implications for the origins of biological compolexity and the missing genetic variation of common diseases. *Philosophical Transactions Royal Society B* **368**: 20120267 (2013).

Germline selection on mitochondrial DNA

Fan W, Waymire KG, Narula N, *et al*. A mouse model of mitochondrial disease reveals germline selection against severe mtDNA mutations. *Science* **319**: 958–62 (2008).

Stewart JB, Freyer C, Elson JL, Wredenberg A, Cansu Z, Trifunovic A, Larsson N-G. Strong purifying selection in transmission of mammalian mitochondrial DNA. *PLoS Biology* **6**: e10 (2008).

Haldane's rule

Coyne JA, Orr HA. Speciation. Sinauer Associates, Sunderland MA (2004).

Haldane JBS. Sex ratio and unisexual sterility in hybrid animals. *Journal of Genetics* **12**: 101–109 (1922).

Johnson NA. Haldane's rule: the heterogametic sex. *Nature Education* **1**: 58 (2008).

Mitochondria and metabolic rate in sex determination

Bogani D, Siggers P, Brixet R *et al*. Loss of mitogen-activated protein kinase kinase kinase 4 (MAP3K4) reveals a requirement for MAPK signalling in mouse sex determination. *PLoS Biology* **7**: e1000196 (2009).

Mittwoch U. Sex determination. *EMBO Reports* **14**: 588–92 (2013).

Mittwoch U. The elusive action of sex-determining genes: mitochondria to the rescue? *Journal of Theoretical Biology* **228**: 359–65 (2004).

Temperature and metabolic rate

Clarke A, Pörtner H-A. Termperature, metabolic power and the evolution of endothermy. *Biological Reviews* **85**: 703–27 (2010).

Mitochondrial diseases

Lane N. Powerhouse of disease. *Nature* **440**: 600–602 (2006).

Schon EA, DiMauro S, Hirano M. Human mitochondrial DNA: roles of inherited and somatic mutations. *Nature Reviews Genetics* **13**: 878–90 (2012).

Wallace DC. A mitochondrial bioenergetic etiology of disease. *Journal of Clinical Investigation* **123**: 1405–12 (2013).

Zeviani M, Carelli V. Mitochondrial disorders. *Current Opinion in Neurology* **20**: 564–71 (2007).

Cytoplasmic male sterility

Chen L, Liu YG. Male sterility and fertility restoration in crops. *Annual Review Plant Biology* **65**: 579–606 (2014).

Innocenti P, Morrow EH, Dowling DK. Experimental evidence supports a sex-specific selective sieve in mitochondrial genome evolution. *Science* **332**: 845–48 (2011).

Sabar M, Gagliardi D, Balk J, Leaver CJ. ORFB is a subunit of F_1F_0-ATP synthase: insight into the basis of cytoplasmic male sterility in sunflower. *EMBO Reports* **4**: 381–86 (2003).

Haldane's rule in birds

Hill GE, Johnson JD. The mitonuclear compatibility hypothesis of sexual selection. *Proceedings Royal Society B* **280**: 20131314 (2013).

Mittwoch U. Phenotypic manifestations during the development of the dominant and default gonads in mammals and birds. *Journal of Experimental Zoology* **281**: 466–71 (1998).

Requirements for flight

Suarez RK. Oxygen and the upper limits to animal design and performance. *Journal of Experimental Biology* **201**: 1065–72 (1998).

The apoptotic death threshold

Lane N. Bioenergetic constraints on the evolution of complex life. *Cold Spring Harbor Perspectives in Biology*. **doi:** 10.1101/cshperspect.a015982 (2014).

Lane N. The costs of breathing. *Science* **334**: 184–85 (2011).

Incidence of early occult miscarriage in humans

Van Blerkom J, Davis PW, Lee J. ATP content of human oocytes and developmental potential and outcome after in-vitro fertilization and embryo transfer. *Human Reproduction* **10**: 415–24 (1995).

Zinaman MJ, O'Connor J, Clegg ED, Selevan SG, Brown CC. Estimates of human fertility and pregnancy loss. *Fertility and Sterility* **65**: 503–509 (1996).

Free radical theory of ageing

Barja G. Updating the mitochondrial free-radical theory of aging: an integrated view, key aspects, and confounding concepts. *Antioxidants and Redox Signalling* **19**: 1420–45 (2013).

Gerschman R, Gilbert DL, Nye SW, Dwyer P, Fenn WO. Oxygen poisoning and X irradiation: a mechanism in common. *Science* **119**: 623–26 (1954).

Harmann D. Aging – a theory based on free-radical and radiation chemistry. *Journal of Gerontology* **11**: 298–300 (1956).

Murphy MP. How mitochondria produce reactive oxygen species. *Biochemical Journal* **417**: 1–13 (2009).

Problems with the free radical theory of ageing

Bjelakovic G, Nikolova D, Gluud LL, Simonetti RG, Gluud C. Antioxidant supplements for prevention of mortality in healthy participants and patients with various diseases. *Cochrane Database of Systematic Reviews* **doi**: 10.1002/14651858.CD007176 (2008).

Gutteridge JMC, Halliwell B. Antioxidants: Molecules, medicines, and myths. *Biochemical Biophysical Research Communications* **393**: 561–64 (2010).

Gnaiger E, Mendez G, Hand SC. High phosphorylation efficiency and depression of uncoupled respiration in mitochondria under hypoxia. *Proceedings National Academy Sciences* **97**: 11080–85 (2000)

Moyer MW. The myth of antioxidants. *Scientific American* **308**: 62–67 (2013).

Free-radical signalling in ageing

Lane N. Mitonuclear match: optimizing fitness and fertility over generations drives ageing within generations. *BioEssays* **33**: 860–69 (2011).

Moreno-Loshuertos R, Acin-Perez R, Fernandez-Silva P, Movilla N, Perez-Martos A, de Cordoba SR, Gallardo ME, Enriquez JA. Differences in reactive oxygen species production explain the phenotypes associated with common mouse mitochondrial DNA variants. *Nature Genetics* **38**: 1261–68 (2006).

Sobek S, Rosa ID, Pommier Y, *et al*. Negative regulation of mitochondrial transcrioption by mitochondrial topoisomerase I. *Nucleic Acids Research* **41**: 9848–57 (2013).

Free radicals in relation to rate of living theory

Barja G. Mitochondrial oxygen consumption and reactive oxygen species production are independently modulated: implications for aging studies. *Rejuvenation Research* **10**: 215–24 (2007).

Boveris A, Chance B. Mitochondrial generation of hydrogen peroxide – general properties and effect of hyperbaric oxygen. *Biochemical Journal* **134**: 707–16 (1973).

Pearl R. *The Rate of Living. Being an Account of some Experimental Studies on the Biology of Life Duration.* University of London Press, London (1928).

Free radicals and the diseases of old age

Desler C, Marcker ML, Singh KK, Rasmussen LJ. The importance of mitochondrial DNA in aging and cancer. *Journal of Aging Research* **2011**: 407536 (2011).

Halliwell B, Gutteridge JMC. *Free Radicals in Biology and Medicine*. 4th edition. Oxford University Press, Oxford (2007).

He Y, Wu J, Dressman DC, *et al*. Heteroplasmic mitochondrial DNA mutations in normal and tumour cells. *Nature* **464**: 610–14 (2010).

Lagouge M, Larsson N-G. The role of mitochondrial DNA mutations and free radicals in disease and ageing. *Journal of Internal Medicine* **273**: 529–43 (2013).

Lane N. A unifying view of aging and disease: the double agent theory. *Journal of Theoretical Biology* **225**: 531–40 (2003).

Moncada S, Higgs AE, Colombo SL. Fulfilling the metabolic requirements for cell proliferation. *Biochemical Journal* **446**: 1–7 (2012).

Aerobic capacity and lifespan

Bennett AF, Ruben JA. Endothermy and activity in vertebrates. *Science* **206**: 649–654 (1979).

Bramble DM, Lieberman DE. Endurance running and the evolution of Homo. *Nature* **432**: 345–52 (2004).

Koch LG Kemi OJ, Qi N, *et al*. Intrinsic aerobic capacity sets a divide for aging and longevity. *Circulation Research* **109**: 1162–72 (2011).

Wisløff U, Najjar SM, Ellingsen O, *et al*. Cardiovascular risk factors emerge after artificial selection for low aerobic capacity. *Science* **307**: 418–420 (2005).

Epilogue: From the Deep

Prokaryote or eukaryote?

Wujek DE. Intracellular bacteria in the blue-green-alga *Pleurocapsa minor*. *Transactions American Microscopical Society* **98**: 143–45 (1979).

Yamaguchi M, Mori Y, Kozuka Y, *et al*. Prokaryote or eukaryote? A unique organism from the deep sea. *Journal of Electron Microscopy* **61**: 423–31 (2012).

LIST OF ILLUSTRATIONS

INDEX

of protein folding 59
relation to free energy 61
and the second law 22
and spores 56–7, 59
environment
explanation for apparent characteristics of
LUCA 130
genome size and 23
importance of membranes and 77
interplay of genes and 29–34
neglected in definitions of life 55
possible influence on mitochondria 264–5
reactivity and energy flux 62–3, 80
enzymes 293
environmental sodium and 144–5
function in metabolism 89–90, 101
metal-sulphide cofactors 100–1
see also ATP synthase; Ech
eocyte hypothesis 166
Epulopiscium 176–9
Errington, Jeff 209
'error catastrophe' proposal 269
Euglena 35
eukaryotes 293–4
archaea similarities with 8–9, 124
archezoa as 48
bacteria similarities with 124
common traits 51, 86, 160, 193, 215
conserved features 41–2, 247
endosymbiosis in phagocytic eukaryotes 181
energy per gene 171
evolution of the nucleus 194–6
excluding *Parakaryon myojinensis* 283–4
extremophile 43
genes as fragmented 199–200
genes derived by endosymbiosis 164, 185
genes derived from archaea and bacteria
162
genetic instability of early forms 197–8,
207–8
lacking mitochondria 37
as monophyletic 40, 49–50

morphological complexity and
morphological similarities 34–6
origins 26, 34, 161–9, 179–85
signature genes 162–3
signature proteins 43
straight chromosomes in 42, 47, 158–60,
217–18
supergroups within 40–1
ubiquity of sex among 42, 199, 215, 220
see also last eukaryotic common ancestor
eukaryotic cells
chimeric origins 12
defined 6
morphological similarity 1, 34–6
serial endosymbiosis theory 6
sizes 158
transport systems within 189–90
evolution
cellular evolution 3, 96–102
of chemiosmotic coupling 83, 85
convergent evolution 20, 45, 151
energetic constraints on 21
of eyes, flight and multicellularity 45–6
gradualism in 194
'by jerks and creeps' 162
as not predictive 21, 221
of protocells 135
reductive evolution 39
timeline 26
see also mutations; natural selection
evolutionary intermediates
absence 2, 44–5, 160, 197
archezoa proposed as 39
genuine examples 48
loss predicted by natural selection 45
evolutionary recapitulation 285–6
exergonic reactions 114, 131, 150, 294
exoplanets, earth-like 287
exothermic reactions 108, 113, 294
extinction risk, asexual reproduction 214–15
extraterrestrial life *see* life elsewhere
extremophiles, eukaryotic 43

stromatolites 26–7
structural constraints on bacterial evolution 158–60
substrates 298
sulphate and sulphite respiration 32, 79
sun, energy received from 64
supercomplexes 238
supergroups, eukaryotic 40–1
superoxide dismutase 246
superoxide radicals 245–6, 269, 271n, 275
surface-to-volume ratios
    enlarging bacteria 173, 185
    vesicle growth and division 99
Sutherland, John 92
symbiosis 5, 36
    see also endosymbiosis
Szent-Györgyi, Albert 28

T
tardigrades 55–6
temperature
    around hydrothermal vents 92n, 106, 109–10
    and free energy change 61
    and hybrid breakdown 258
    and sex determination 253–4
terminally differentiated tissues 256
terminology 15, 49n
testes 254, 255n, 260, 276
Thauer, Rolf 150
thermodynamics 298
    interplay with kinetics 113–14, 298
    second law 22, 56, 96
thermophoresis 110–12, 119, 134, 299
thioesters 79, 133–4, 148
Thiomargarita 176–8
The Third Man (film) 157
Tigriopus californicus 249
Tiktaalik 48
timeline of evolution 26
tissues
    dysfunctional 228–30

metabolic requirements 255–6, 260n
    non-regenerating 232, 256
topoisomerase-1 271
trade-offs *see* cost-benefit
traits
    common to all complex life 1, 14
    common to all eukaryotes 51, 86, 160, 193, 215
transcription and translation 299
transcription factors 144, 204n, 271n
transport systems in eukaryotic cells 189–90
trees of life
    'amazing disappearing tree' 126–7
    archezoa in 38
    based on informational genes 166
    Bill Martin's version 12, 126–8, 164–5
    Carl Woese and 7–8
    Darwin and 122
    eukaryotic supergroups 40–1
    lateral gene transfer and 123–6, 128
    Lynn Margulis and 9
    single gene and full genome 161
    see also phylogenetics
Tribolium spp. 258
trivial hypothesis 186
two sexes 43, 219–26, 241

U
ubiquinone 67, 72
unikonts 40–1
uniparental inheritance 299
    advantages and disadvantage 223–4
    of mitochondria 220–3
    origins 230–1
    universality in multicellular animals 225, 231
    variance increased by 222, 225, 227
units
    ångströms and nanometres 69n
    of entropy 56
universe, life elsewhere *see* life elsewhere
Urey, Harold 90–1